CONTENTS

Foreword by the Translator vii
Foreword to the U. S. Edition ix
Foreword to the 1937 Edition xiii

Part One

THE WAR OF MOVEMENT

BELGIUM AND NORTHERN FRANCE, 1914

Chapter 1: Fighting at Bleid and Doulcon Woods
I: The Beginning, 1914—Ulm, July 31, 1914 1
II: At the Frontier 2
III: Reconnaissance in the Direction of Longwy and Preparations for the First Battle 3
IV: The Battle of Bleid 6
V: On the Meuse; Battles at Mont and in the Doulcon Woods 14

Chapter 2: Battles at Gesnes, Defuy Woods and Rembercourt
I: The Fight at Gesnes 22
II: Pursuit Through the Argonne; The Fight at Pretz 24
III: Attack on Defuy Woods 26
IV: Battle at the Defuy Woods 30
V: Night Attack From September 9-10, 1914 35

Chapter 3: Fighting Near Montblainville
I: Retirement Through the Argonne 40
II: Engagement at Montblainville; Storming Bouzon Woods 42
III: Forest Fighting Along the Roman Road 46

Part Two

TRENCH WARFARE

IN THE ARGONNE AND HIGH VOSGES

Chapter 4: Attack in the Charlotte Valley 51

Chapter 5: Trench Fighting at "Central" and in the Charlotte Valley
I: Trench Warfare in the Argonne 62
II: Attack on Central 65
III: Attack of September 8, 1915 71

Chapter 6: Raids in the "Pinetree Knob" Sector, High Vosges
I: The New Unit 75
II: Raids in the "Pinetree Knob" Sector 76

CONTENTS—*Continued*

Part Three

OPEN WARFARE IN RUMANIA AND THE CARPATHIANS, 1917

Chapter 7: From Skurduk Pass to Vidra
I: Occupation of Hill 1794 82
II: Attack on the Lesului 85
III: Battle at Kurpenul—Valarii 88
IV: Hill 1001, Magura Odobesti 95
V: Gagesti 101
VI: At Vidra 109

Chapter 8: First Operations Against Mount Cosna
I: Approach March to the Carpathian Front 112
II: Attack against the Ridge Road Salient, August 9, 1917 114
III: Attack of August 10, 1917 122
IV: The Capture of Mount Cosna, August 11, 1917 129
V: Combat on August 12, 1917 140

Chapter 9: Further Operations at Mount Cosna
I: The Defense, August 14-18, 1917 145
II: Second Attack on Mount Cosna, August 19, 1917 157
III: Again on the Defensive 165

Chapter 10: The First Day of the Tolmein Offensive
I: Approach March and Preparation for the Twelfth Battle of the Isonzo 168
II: The First Attack: Hevnik and Hill 1114 172

Chapter 11: The Second Day of the Tolmein Offensive
I: Surprise Breakthrough to the Kolovrat Position 185
II: Attack against Kuk. The Barring Luico-Savogna Valley and Opening of the Luico Pass 194

Chapter 12: The Third Day of the Tolmein Offensive
I: The Assault on Mount Cragonza 208
II: The Capture of Hill 1192 and the Mrzli Peak (1356) and the Attack on Mount Matajur 218

Chapter 13: Pursuit Across the Tagliamento and Piave Rivers, October 26, 1917—January 1, 1918
I: Masseris—Campeglio—Torre River—Tagliamento River—Klautana Pass 228
II: Pursuit to Cimolais 233
III: Attack Against the Italian Positions West of Cimolais 235
IV: Pursuit Through Erto and Vajont Ravine 240
V: The Fight at Longarone 243
VI: Battles in the Vicinity of Mount Grappa 260

Foreword by the Translator

This book is a translation of Lieutenant Colonel (now General Field Marshal) Erwin Rommel's *Infanterie Greift an,* which was published in Germany in 1937. It is written in a style familiar to readers of *Infantry in Battle,* with the difference that it is a continuous narrative rather than a compilation of separate sections.

We have transposed all German units and ranks into their American equivalents. Likewise all measurements have been converted from the metric system. The only exceptions are the designation of heights and here the equivalent in feet has been inserted parenthetically.

This translation was not prepared with the author's sanction, and the translator has been obliged to devise his own clarification of certain questionable points.

The main theme emphasizes the importance of the basic principles of training, of security, of prior planning, and of the vital need for initiative and hard work on the part of all junior commanders.

GUSTAVE E. KIDDÉ
Lieutenant Colonel
Coast Artillery Corps

Command & General Staff School
Fort Leavenworth, Kansas
July 1943

Foreword to the U. S. Edition

First-rate small-unit combat narratives are useful in training troops and leaders but are not very plentiful. *The Infantry Journal* has published a number of "little picture" battle studies in the past. Notable among these was *Infantry in Battle.* Certainly we should miss no opportunity to learn from our enemies. The Germans made tremendous efforts to find out why they lost the First World War. Hundreds of books were written analyzing and weighing the experiences of 1914-18. One such book, *Infanterie Greift an (Infantry Attacks),* devoted to the experiences of a Württemberg mountain infantry battalion, was written in 1937 by an unknown German officer named Erwin Rommel. He was then a lieutenant colonel completing a tour of duty as instructor in infantry tactics at the Dresden Military Academy. Two years earlier he had written a small handbook for platoon and company leaders entitled, *Aufgaben für Zug und Kompanie (Problems for the Platoon and Company).* Neither of Rommel's books made much of an impression at the time; they were given only perfunctory reviews in German military periodicals, and only a bare mention in British or American military journals.

Five years later Rommel was directing the Afrika Korps with such success that, according to the Gallup Poll even the British, up until November 1942, considered him the "ablest commander produced by the war." His repeated victories in desert operations against a succession of British commanders caused him to become the most publicized German general. His books, which up to 1941 had sold only a few thousand copies, went through many editions in Germany. When our Army wanted to know what Rommel had written in 1937, we found it difficult to lay our hands on *Infanterie Greift an,* although there were a few stray copies in this country. Through the kindness of the Columbia University Library, Colonel Kiddé was able to make this translation for the Command & General Staff School. It is being published by *The Infantry Journal* in order to give it as wide a circulation as possible. Most of the general tactical lessons taught by these combat narratives are valid today. The observations under which Rommel sums up his reactions to the various engagements are precisely the kind of counsel an American officer would give his troops and junior officers under similiar circumstances.

With the memories of Rommel's spectacular campaigns in North Africa still fresh in mind, present-day readers of *Infantry Attacks* will see many parallels between Rommel's experiences and methods in the First and Second World Wars.

As a leader of a small unit in 1914-18, Rommel proved himself to be an aggressive and versatile commander. He had a highly developed capacity for utilizing terrain. His men were trained to take cover when possible in movement and to dig in whenever they stopped. Rommel was tireless in reconnaissance and attributed many of his successes to the fact that he possessed better information about the enemy than they did about him. Information was shared with junior officers, noncoms, and even private soldiers. Into every battle plan and maneuver Rommel tried to introduce some element of deception and surprise. He sought out the weakest element in the enemy position and worked out a plan of attack to exploit that weakness and confuse the enemy as to his real intentions. He took pains to insure proper fire plans and used his machine guns and hand grenades in 1916-18 with the same skill that he used his 88s in 1941-42. Rommel was not afraid of changing plans or disobeying an order if he had better local information than his superior officer. He was also good at judging the moment when the cracking enemy should be attacked with every man at his disposal. If necessary he would order his men into the zone of a German barrage in order to give the enemy no rest in retreat. He bluffed Italians and lied to Rumanians in order to get them to surrender in 1917-18, just as he lied to his own troops in November 1941 (saying that Moscow had fallen) in order to get them to make a supreme effort against General Ritchie's offensive.

The swiftness with which the Afrika Korps switched from armored attack to antitank defense showed that he remembered the lessons of 1914-18. He was constantly making personal reconnaissances in North Africa by station wagon, armored car, or Storch observation plane. His troops called him "the General of the Highway." Instead of sharing his information before battle with subordinates as he did in 1915-18, Rommel broadcast in the clear his instructions and orders by radio in 1942, making use of a map reference called the "thrust line" which enabled him to direct tanks, planes, and motorized infantry amid the fluid conditions of battle. British radio listeners in Lybia often heard Rommel's cool voice directing operations, although without knowing the "thrust line" on which his orders were based they could not understand what he meant or take counter action until too late.

In 1941-42, acting without air superiority, Rommel repeatedly destroyed British tank units larger than his own by striking them in detail. The Afrika Korps dug in its men and guns and set out its minefields, with astonishing swiftness. It prepared fire plans with great care. This enabled Rommel to lure the bulk of General Ritchie's armor into a tank ambush at Knightsbridge Box on June 13, 1942, where he destroyed most of it.

For his victories at Knightsbridge and Tobruk in June 1942, he was awarded the rank of Field Marshal. Until his forward momentum was checked in July 1942 at El Alamein, it looked as if Rommel's deception, speed, and striking power might be too much for the British in the Western Desert.

The arrival of General Montgomery changed all that. He made a new army out of the British Eighth Army by discipline and training. The arrival of new tanks, guns, and self-propelled weapons turned the scales well against the Germans. Rommel was decisively defeated at El Alamein, driven into a retreat which led across Egypt, Cyrenaica, Lybia, Tripoli, into southern Tunisia. Failing to prevent Montgomery from crossing the Mareth line in March-April 1943, Rommel was recalled to Germany for reasons of health. His successor, Colonel General von Arnim surrendered with the Afrika Korps in the Tunisian débâcle of May 6-13, 1943.

When Italy collapsed in September 1943, Rommel was placed in command of the Italian and Balkan fronts. Early in 1944 Rommel was placed in charge of the anti-invasion forces in Western Europe. Despite repeated German references to his poor health, Rommel may again prove to be a resourceful and intrepid leader in battle. No commander can afford to take the slightest chance when fighting against Rommel, or offer him even the suggestion of an advantage. He is a tough and resourceful leader, but as General Montgomery has twice clearly proved, he *can be* outgeneraled and outfought.

MAJOR H. A. DE WEERD
Associate Editor
The Infantry Journal

Washington
February 1944

Foreword to the 1937 Edition

This book describes numerous World War I battles which I experienced as an Infantry officer. Remarks are appended to many descriptions in order to extract worthwhile lessons from the particular operation.

The notes, made directly after combat, will show German youth capable of bearing arms, the unbounded spirit of self-sacrifice and courage with which the German soldier, especially the Infantryman, fought for Germany during the four-and-a-half-year war. The following examples are proof of the tremendous combat powers of the German infantry, even when faced with superior odds in men and equipment; and these sketches are again proof of the superiority of the junior German commander to his enemy counterpart.

Finally, this book should make a contribution towards perpetuating those experiences of the bitter war years; experiences often gained at the cost of great deprivations and bitter sacrifice.

ERWIN ROMMEL
Lieutenant Colonel

INFANTRY ATTACKS

This book was first published by Ludwig Voggenreiter Verlag, Potsdam, Germany, in 1937, under the title: *Infanterie Greift an: Erlebnisse und Erfahrungen.*

Part One

THE WAR OF MOVEMENT
BELGIUM AND NORTHERN FRANCE
1914

★ ★ ★ ★ ★ ★ ★ ★ ★ ★ ★ ★ ★ ★ ★

Chapter 1

FIGHTING AT BLEID AND DOULCON WOODS

I: The beginning, 1914—Ulm, July 31, 1914

The danger of war hung ominously over the German nation. Everywhere, serious, troubled faces! Unbelievable rumors which spread with the greatest of rapidity filled the air. Since dawn all public bulletin boards had been surrounded. One extra edition of the papers followed the other.

At an early hour the 4th Battery of the 49th Field Artillery Regiment hurried through the old imperial city. *Die Wacht am Rhein* resounded in the narrow streets.

I rode as an infantry lieutenant and platoon commander in the smart Fuchs Battery to which I had been assigned since March. We trotted along in the bright morning sunshine, did our normal exercises, and then returned to our quarters accompanied by an enthusiastic crowd whose numbers ran into thousands.

During the afternoon, while horses were being purchased in the barrack yard, I obtained relief from my assignment. Since the situation appeared most serious, I longed for my own regiment, the King Wilhelm I, to be back with the men whose last two years of training I had supervised in the 7th Company, 124th Infantry (6th Württemberger).

Along with Private Hänle, I hurriedly packed my belongings; and late in the evening we reached Weingarten, our garrison city.

On August 1, 1914, there was much activity in the regimental barracks, the big, old cloister building in Weingarten. Field equipment was being tried on! I reported back to headquarters and greeted the men of the 7th Company whom I was to accompany into the field. All the young faces radiated joy, animation, and anticipation. Is there anything finer than marching against an enemy at the head of such soldiers?

At 1800, regimental inspection. Colonel Haas followed his thorough inspection of the field-gray-clad regiment with a vigorous talk. Just as we fell out, the mobilization order came. Now the decision had been made. The shout of German youths eager for battle rang through the ancient, gray cloister buildings.

The 2d of August, a portentous Sabbath! Regimental divine services were held in the bright sunlight, and in the evening the proud 6th Württemberger Regiment marched out to resounding band music and entrained for Ravensburg. An unending stream of troop trains rolled westward toward the threatened frontier. The regiment left at dusk to the accompaniment of cheers. To my great disappointment I was obliged to remain behind for a few days in order to bring up our reserves. I feared that I was going to miss the first fight.

The trip to the front on August 5, through the beautiful valleys and dells of our native land and amid the cheers of our people, was indescribably beautiful. The troops sang and at every stop were showered with fruit, chocolate, and rolls. Passing through Kornwestheim, I saw my family for a few brief moments.

We crossed the Rhine during the night. Searchlights crisscrossed the sky on the lookout for enemy planes or dirigibles. Our songs had died down. The soldiers slept in all positions. I rode in the locomotive, looking now into the firebox then out into the rustling, whispering, sultry summer's night and wondering what the next few days would bring.

In the evening of August 6 we arrived at Königsmachern near Diedenhofen and were glad to be out of the cramped quarters of the troop train. We marched through Diedenhofen to Ruxweiler. Diedenhofen was not a pretty sight with its dirty streets, houses, and taciturn people. It seemed so different from my home in Swabia.

We continued the march, and at nightfall a torrential downpour set in. Soon there was not a dry stitch of clothing on our bodies, and the water-soaked packs began to weigh heavily. A fine beginning! Occasional shots were heard far in the distance. About midnight our platoon arrived in Ruxweiler without suffering any losses during the six-hour march. The company commander, First Lieutenant Bammert, awaited us. Cramped quarters on straw was our lot.

II. At The Frontier

During the next few days, hard drilling welded our war-strength company together. Besides platoon and company exercises, we were subjected to a wide variety of combat exercises which all placed great emphasis on the use of the spade. In addition, I spent several uneventful rainy days on guard with my platoon in the vicinity of Bollingen. Here some of my men and I suffered stomach disturbances as a result of the greasy food and the freshly baked bread.

On August 18 we began our main advance toward the north. I rode my company commander's second mount. Singing gaily, we crossed the German—Luxembourg frontier. The people were friendly and brought fruit and drink for the marching troops. We entered Budersberg.

Early on August 19 we moved to the southwest, passed under the cannon of the French fortress at Longwy, and bivouacked at Dahlem. The first battle was near. My stomach gave me a great deal of trouble, and even a chocolate and zwieback diet brought no relief. I would not report sick for I did not want to be looked upon as a shirker.

On August 20 after a hot march we reached Meix-la-Tige in Belgium. The 1st Battalion garrisoned the outpost line and the 2d Battalion provided local security. The population was very reserved and reticent. A few enemy planes appeared and were fired on without result.

III: Reconnaissance in the Direction of Longwy and Preparations for the First Battle

The next day was to be a day of rest. In the early hours of the morning, several fellow officers and I reported to Colonel Haas who ordered each of us to take a five-man reconnaissance detachment past Barancy and Gorcy in the direction of Cosnes near Longwy to ascertain the enemy dispositions and strength. The distance out was eight miles, and to save time we obtained permission to go by wagon as far as the outpost. Our Belgian drayhorse ran away while we were still in Meix-la-Tige, and the upshot was a landing in a manure pile. With only a broken-down wagon to show as a result of our efforts, we continued our way on foot.

Burdened with the responsibility of human life, we moved forward with a greater degree of caution than was normal in peacetime maneuvers. We left the town by means of a ditch along the side of the road. The road wound through grain fields on the way to Barancy which had been reported on the previous day as being occupied by weak enemy forces. On arriving we found it unoccupied; and leaving the highway and passing through grain fields, we crossed the Franco-Belgian frontier, reached the southern edge of the Bois de Mousson, and then descended towards Gorcy. The detachment under Lieutenant Kirn followed us, covering our movement through Gorcy from a hilltop.

On the Gorcy-Cosnes highway, we found signs that enemy infantry and cavalry were moving in the direction of Cosnes. Greater caution was indicated; we moved off the road and continued our march through the heavy growth bordering the road. Maintaining careful observation of the road, we finally reached a clump of woods five hundred yards west of Cosnes. I studied the terrain with field glasses but saw no French troops. On our way across the open fields to Cosnes, we came upon an old woman peacefully at work. She related in German that the French troops had left Cosnes for Longwy an hour before and that no other troops remained in Cosnes. Would the old woman's story hold water?

We worked our way through grain fields and orchards and entered Cosnes with fixed bayonets, fingers on triggers, and all eyes studying doorways and windows for telltale evidence of an ambush. However, the inhabitants appeared friendly and confirmed the old woman's statement. They brought us food and drink, but we were still distrustful and made them sample the food before helping ourselves.

To speed reporting I seized six bicycles giving quartermaster receipts in return. Using our newly acquired conveyances, we pedaled a mile down the road in the direction of Longwy on whose outer works heavy artillery fire was being laid. Far and wide, nothing was to be seen of enemy troops. The mission of the reconnaissance detachment had now been accomplished. At a fast clip we passed through Gorcy on our way down grade to Barancy. We maintained a considerable interval between men and carried our guns ready for use under our arms. From Barancy on, I went on ahead of my men in order to report quickly.

On the street of Meix-la-Tige, I met the regimental commander and made my report. Tired and hungry, I headed for my quarters, looking forward to a few hours' rest. No such luck. In front of the quarters my battalion was drawn up ready to move. Hänle, efficient as usual, had already packed my belongings and saddled my horse. Before shoving off there was not even enough time for a bite to eat.

We marched to a hill three-quarters of a mile southeast of Saint Léger. The sky was overcast. From the southwest came the sound of rifle and occasional artillery fire. We knew that elements of the 1st Battalion, which were still on outpost duty near Villancourt, had made contact with the enemy during the afternoon.

At nightfall the regiment, less the 1st Battalion, went into bivouac some two miles south of Saint Léger with our security elements about three-quarters of a mile ahead. I was getting ready for a night's sleep when a call came for me to report to the regimental CP located some fifty yards from my platoon bivouac area. Colonel Haas asked whether I would make a trip through the woods to the 1st Battalion at Villancourt. My mission was to give the 1st Battalion the regimental order to retire to Hill 312 by the shortest route possible, and I was appointed battalion guide. (See sketch 1.)

With Sergeant Gölz and two men from the 7th Company, I went on my way. We traveled in the dark by compass through the meadowland southeast of Hill 312. Off to the right we heard our own sentries' challenges, now and then a rifle shot. Soon we were climbing a steep, thickly-wooded slope. From time to time we halted and listened to the noises of the night. Finally, after a hard climb and feeling our way, we reached the crest of the line of hills west of Villancourt.

To the southeast we could see the glow from Longwy fortress which had been set on fire as a result of the artillery bombardment. We descended through the thick brush toward Villancourt. Suddenly from close at hand a sentry called out: "Halt, who is there?" Was he German or French? We knew that the French often challenged in German. We dropped to the ground. "Give the countersign!" None of us knew it. I called my name and rank—and was recognized. Some 1st Battalion outposts were located on the edge of the woods.

It was not much farther to Villancourt. Five hundred yards south of the town we found companies of the 1st Battalion resting on the side of the Villancourt—Mussy-la-Ville road in close order.

I transmitted the regimental order to the battalion commander, Major Kaufmann. Compliance was not possible, for the 1st Battalion was still attached to the Langer Brigade. I was taken to General Langer's CP, on the hill one-half mile southwest of Villancourt, to give him my message. General Langer ordered me to return to my regiment with the information that he could not spare our 1st Battalion until the remainder of his brigade came up to Villancourt. Downcast at the failure of our mission and physically exhausted, my three companions and I headed back to Hill 312.

It was past midnight when I arrived at the regimental CP. I woke the regimental adjutant, Captain Volter, and reported. Colonel Haas also heard it. He was not greatly pleased and ordered me to go by a round-about way to the 53d Brigade at Saint Léger, either on foot or mounted, and report personally to the brigade commander, General von Moser, that General Langer would not release the 1st Battalion, 124th Infantry. Did I tell my colonel that this job was beyond my strength, that I had been on the go for eighteen hours and was now exhausted? No; although a tough job lay ahead, it had to be done.

I groped my way to the company commander's second mount, tightened the girth and rode off to the north. I found General von Moser in a tent on the hill a short distance southeast of Saint Léger. He was extremely displeased at my report and ordered me to return to Villancourt by way of the regimental CP and inform General Langer that the 1st Battalion of the 124th Regiment had to be under regimental control by daybreak.

I covered a total distance of six miles, part of it on horse and part on foot, delivered my message and got back to Hill 312 as dawn was breaking. All units were ready, rations had been issued and eaten, and the kitchens had pulled out. My orderly, Hänle, helped me out with a swig from his canteen. Dense, wet fog surrounded us. At the regimental CP, orders were being issued.

Observations: Facing the enemy, the reconnaissance detachment commander becomes conscious of his heavy responsibilities. Every mistake means casualties, perhaps the lives of his men. Therefore any advance must be made with extreme caution and deliberation. Taking advantage of all cover, the detachment should keep off the roads and repeatedly examine the terrain with field glasses. The detachment should be organized in considerable depth. Before crossing open stretches of terrain fire support must be arranged for. In entering a village, advance with part of the unit on the left, the rest on the right of the houses and with fingers on the triggers. Report observations rapidly, for delay lessens the value of any information.

Train in time of peace to maintain direction at night with the aid of a luminous dial compass. Train in difficult, trackless, wooded terrain. War makes extremely heavy demands on the soldier's strength and nerves. For this reason make heavy demands on your men in peacetime exercises.

IV: The Battle of Bleid

About 0500, the 2d Battalion started off for Hill 325 about a mile and a half northeast of Bleid. A thick ground fog lay on the dew-covered fields, limiting visibility to a scant fifty yards. The battalion commander, Major Bader, sent me on ahead to explore the road to Hill 325. Having been on the go for nearly twenty-four hours, I could scarcely stay in the saddle. The terrain on both sides of the country road over which I rode was covered with numerous hedges and fenced-in meadows. With map and compass I found Hill 325; the battalion came up and deployed on the northeast slope.

Soon afterward our advanced security elements on the south and west slope of Hill 325 ran into the enemy in the fog. A brief exchange of shots was heard from several directions. Occasional rifle bullets whined above our heads; what a peculiar sound! An officer who had ridden a few hundred yards in the direction of the enemy was fired on from close range. Riflemen rushed forward and succeeded in bringing down a red-trousered Frenchman and took him prisoner.

Now we heard German commands off to the left and toward the rear: "Half left, march! Increase distances!" A skirmish line suddenly emerged from the fog. It was the right wing of the 1st Battalion. My company commander ordered me to deploy my platoon, make contact with the right of the 1st Battalion, and advance on the southeast of Bleid.

I turned my horse over to Hänle, exchanged my automatic for his bayonet, and deployed my platoon. In skirmish formation we advanced toward Bleid through potato fields and vegetable gardens over the southeast slope of Hill 325. A heavy fog hung over the fields and visibility was still limited to fifty or eighty yards.

Suddenly a volley was fired at us from close range. We hit the dirt and lay concealed among the potato vines. Later volleys passed high over our heads. I searched the terrain with my glasses but found no enemy. Since he obviously could not be far away, I rushed toward him with the platoon. But the French got away before we had a chance to see him, leaving clearly defined tracks in the potato field. We continued on toward Bleid. In the excitement of the fight, we lost contact with the right wing of the 1st Battalion.

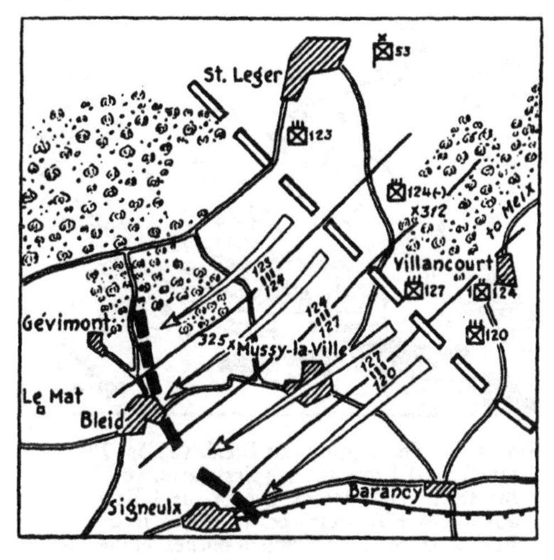

Sketch 1 : The Fighting at Bleid

Several additional volleys were fired at the platoon from out of the fog; but each time we charged, the enemy withdrew hastily. We then proceeded about a half mile without further trouble. Suddenly a high hedged fence appeared through the fog, and to the right rear we saw the outlines of a farm. At the same time, we began to distinguish a group of tall trees to the left. The footprints of the enemy we had been following turned off to the right and went up the slope. Was Bleid in front of us? I left the platoon in the shelter of the hedge and sent out a scouting detachment to make contact with our neighbors on the left and with our own outfit. So far the platoon had suffered no casualties.

I went on ahead with Sergeant Ostertag and two range estimators to investigate the farm ahead of us. Nothing could be seen or heard of the enemy. We reached the east side of the building and found a narrow dirt path leading down to a highway on the left. On the far side through the fog we could distinguish another group of farm buildings. Without doubt we were on the Mussy-la-Ville side of Bleid. Cautiously we approached the highway; I peered around the corner of the building. There! scarcely twenty paces to the right I saw fifteen or twenty Frenchmen standing in the middle of the highway drinking coffee, chatting, their rifles lying idly in their arms. They did not see me. (These troops were part of the 5th Company of the French 101st Infantry Regiment who were to take up defensive positions at the southeast exit of Bleid.)

I withdrew quickly behind the building. Was I to bring up the platoon? No! Four of us would be able to handle this situation. I quickly informed my men of my intention to open fire. We quietly released the safety catches; jumped out from behind the building; and standing erect, opened fire on the enemy nearby. Some were killed or wounded on the spot; but the majority took cover behind steps, garden walls, and wood piles and returned our fire. Thus, at very close range, a very hot fire fight developed. I stood taking aim alongside a pile of wood. My adversary was twenty yards ahead of me, well covered, behind the steps of a house. Only part of his head was showing. We both aimed and fired almost at the same time and missed. His shot just missed my ear. I had to load fast, aim calmly and quickly, and hold my aim. That was not easy at twenty yards with the sights set for 440 yards, especially since we had not practiced this type of fighting in peacetime. My rifle cracked; the enemy's head fell forward on the step. There were still about ten Frenchmen against us, a few of whom were completely covered. I signalled to my men to rush them. With a yell we dashed down the village street. At this moment Frenchmen suddenly appeared at all doors and windows and opened fire. Their superiority was too much; we withdrew as fast as we had advanced and arrived without loss at the hedge where our platoon was getting ready to come to our aid. Since this was no longer necessary, I ordered everyone back under cover. We were still being fired on through the fog from a building on the far side of the street, but the fire was high. Using my field glasses, I managed to locate the target which was some seventy yards away and I found that the enemy was firing from the roof as well as from the ground floor of a farmhouse. A number of rifle barrels were protruding from the roof

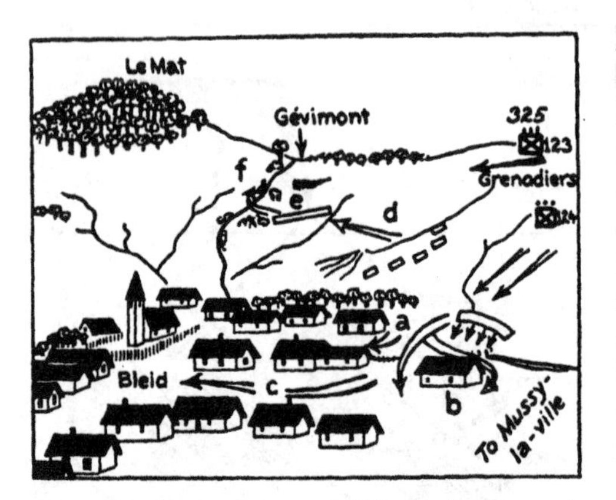

Sketch 2: The Fighting at Bleid. (a) 1st Platoon's attack. (b) Storming the first farm. (c) Fight in the town. (d) Attack across the hills north of Bleid. (e) Fire attack on the enemy in the wheat-field. Seizure of the clump of bushes on the Bleid—Gévimont Road.

tiles. Since it was impossible for the enemy to employ both rear and front sights in firing in this manner, this must have accounted for his fire going high over our heads.

Should I wait until other forces came up or storm the entrance of Bleid with my platoon? The latter course of action seemed proper.

The strongest enemy force was in the building on the far side of the road. Therefore we had to take this building first. My attack plan was to open fire on the enemy on the ground floor and garret of the building with the 2d Section and go around the building to the right with the 1st Section and take it by assault.

Quickly the assault detachment picked up a few timbers which were lying close at hand. These were just the thing for battering down doors and gates. We also took a few bunches of straw along in order to smoke out any concealed men. Meanwhile the 2d Section had been lying along the hedge, ready to fire. The assault detachment had made its preparations under perfect cover. We were ready to start.

On signal, the 2d Section opened fire. I dashed forward to the right with the 1st Section—over the same route I had passed over a few minutes before with the platoon—across the street. The enemy in the house opened with heavy rifle fire mainly directed at the section behind the hedge. The assault detachment was now sheltered by the building and safe from the hostile fire. The doors gave way with a crash under heavy blows of the battering ram. Burning bunches of straw were thrown onto the threshing floor, which was covered with grain and fodder. The building had been surrounded. Anyone who had taken a notion to leap out would have landed on our bayonets. Soon bright flames leapt from the roof. Those of the enemy who were still alive laid down their arms. Our casualties consisted of a few slightly wounded.

We now rushed from building to building. The 2d Section was called up. Wherever we ran into the enemy, he either surrendered or took cover in the building recesses from which he was soon routed. Other elements of the 2d Battalion which had mingled with those of the 1st Battalion now forced their way through the entire village, which was afire in many places. The formations became intermingled. Rifle fire came from all directions and casualties mounted.

In a side street I rushed forward to a church surrounded by a wall from which heavy rifle fire was being directed at us. Making use of available cover and rushing from house to house, we approached the enemy. As we advanced to the assault, he gave way, retreated westward, and was soon lost in the fog.

We now received very heavy fire on our left flank from the south part of Bleid, and our casualties began to increase. On every side we heard the piteous cry for medical help. An aid station was established behind the laundry. Most of the wounds were severe. Some of the men cried with pain; others looked death in the eye with the composure of heroes.

In the northwest and south portions of Bleid the French were still in possession. Behind us the town was ablaze. In the meantime the sun had dissipated the fog. Nothing more could now be done in Bleid; so I assembled everyone within reach, arranged stretcher parties for the wounded, and moved off toward the northeast. I wanted to get out of this cauldron and reestablish contact with my own outfit. Fire; dense, stifling smoke; glowing timbers, crumbling houses; and frightened cattle running wildly among the burning buildings barred our way. Finally, half suffocated, we reached the open. First we took care of the many wounded; then I assembled the formation of about one hundred men and headed on to the shallow depression three hundred yards northeast of Bleid. There I left the platoon, deployed to the west, and went with the section leaders on reconnaissance to the next rise in the terrain. (See sketch 2.)

To the right and above us lay Hill 325 still covered with fog. In the tall fields of grain on its southern slope, we could not recognize friend or foe. Off to the right and about half a mile ahead of us on the far side of a draw, we saw the red breeches of French infantry in company strength on the front edge of a yellow wheatfield behind fresh earthworks. (They belonged to the 7th Company of the French 101st Infantry Regiment.) In the low area to the left and below us, the fight for burning Bleid still raged. Where were our company and the 2d Battalion? Were some still in Bleid with their bulk farther to the rear? What was I to do? Since I did not wish to remain idle with my platoon, I decided to attack the enemy opposite us in the sector of the 2d Battalion. Our deployment behind the ridge, our movement into position, and the opening of fire by the platoon was carried out with the composure and precision of a peacetime maneuver. Soon the groups were in echelon, part of them in the potato field, part of them well concealed behind the bundles of oats from whence they delivered a slow and well-aimed fire as they had been taught to do in peacetime training.

As soon as the leading squads went into position, the enemy opened with heavy rifle fire. But his fire was still too high. Only a few bullets struck in front of and beside us, and we soon became accustomed to this. The only result of fifteen minutes' fire was a hole in a messkit. Half a mile to our rear we saw our own skirmish line advancing over Hill 325. This assured support for our right, and the platoon was now free to attack. We rushed forward by groups, each being mutually supported by the others, a maneuver we had practiced frequently during peacetime. We crossed a depression which was defiladed from the enemy's fire. Soon I had nearly the whole platoon together in the dead angle on the opposite slope. Thanks to poor enemy marksmanship, we had suffered no casualties up to this time. With fixed bayonets, we worked our way up the rise and to within storming distance of the hostile position. During this movement the enemy's fire did not trouble us, for it passed high over us toward those portions of the platoon that were still a considerable distance behind us. Suddenly, the enemy's fire ceased entirely. Wondering if he was preparing to rush us, we assaulted his position but, except for a few dead, found it deserted. The tracks of the enemy led off to the west through the field in which the grain was as tall as a man. Again I found myself well in advance of my own line with my platoon.

I decided to wait until our neighbors on the right came up. The platoon occupied the position they had just gained; then, together with the commander of the 1st Section, a first sergeant of the 6th Company, and Sergeant Bentele, I went off on reconnaissance to the west to learn where the enemy had gone. The platoon maintained contact. Some four hundred yards north of Bleid we reached the road connecting Gévimont and Bleid without having encountered the enemy. The road became higher as it went to the north, passing through a cut at this point. On both sides of the road large clumps of bushes interfered with the view to the northwest and west. We used one of these clumps of bushes as an OP. Strange to say, nothing was to be seen of the retreating enemy. Suddenly, Bentele pointed with his arm to the right (north). Scarcely 150 yards away the grain was moving; and through it we saw the sun's reflection on bright cooking gear piled on top of the tall French packs. The enemy was withdrawing from the fire of our guns which were sweeping the highest portion of the ridge to the west from Hill 325. I estimated that about a hundred Frenchmen were coming straight at us in column of files. Not one of them lifted his head above the grain. (These soldiers belonged to the 6th Company of the French 101st Infantry Regiment. They had been attacked on the west slope of Hill 325 by elements of the 123d Grenadier Regiment and were now retreating toward the southwest.)

Was I to call up the remainder of the platoon? No! They could give us better support from their present position. The penetration effect of our rifle ammunition came to mind! Two or three men at this distance! I fired quickly at the head of the column from a standing position. The column dispersed into the field; then, after a few moments, it continued the march in the same direction and in the same formation. Not a single Frenchman raised his head to locate this new enemy who had appeared so suddenly and so close to him. Now the three of us fired at the same time. Again the column disappeared for a short time, then split into several parts and hastily dispersed in a westerly direction toward the Gévimont-Bleid highway. We opened with rapid fire on the fleeing enemy. Strange to say, we had not been fired on even though we were standing upright and were plainly visible to the enemy. To the left, on the far side of the clump of bushes where we were standing, Frenchmen came running down the highway. They were easily shot down as we fired at them through a break in the bushes at a range of about ten yards. We divided our fire and dozens of Frenchmen were put out of action by the fire of our three rifles.

The 123d Grenadier Regiment was advancing up the slope to the right. I signalled my platoon to follow, and we then advanced northwards on both sides of the Gévimont-Bleid road. During our advance we encountered a number of Frenchmen in the bushes along the road. It took a lot of talking to get them out of their hiding places and make them lay down their arms. They had been taught that the Germans would behead all their prisoners. We got more than fifty men out of the bushes and grain fields, including two French officers, a captain and a lieutenant who had been slightly wounded in the arm. My men offered the prisoners cigarettes which increased their confidence.

To the right on the hill the 123d Grenadier Regiment also reached the Gévimont-Bleid road. We were being fired on from the direction of the forest-covered peak, Le Mat, which was five thousand feet high and lay northwest of Bleid. As quickly as possible I got the platoon into the cut on the right so they would be under cover, with the intention of resuming the fight with an attack on Le Mat from this point. Suddenly, however, everything went black before my eyes and I passed out. The exertions of the previous day and night; the battle for Bleid and for the hill to the north; and, last but not least, the terrible condition of my stomach had sapped the last ounce of my strength.

I must have been unconscious for some time. When I came to, Sergeant Bentele was working over me. French shell and shrapnel were striking intermittently in the vicinity. Our own infantry was retiring toward Hill 325 from the direction of the Le Mat woods. What was it, a retreat? I commandeered part of a line of riflemen, occupied the slope along the Gévimont-Bleid road, and ordered them to dig in. From the men I learned that they had sustained heavy casualties in Le Mat woods, had lost their commander, and that their withdrawal was executed on orders from a superior commander. Above all, French artillery wrought great havoc among them. A quarter of an hour later, buglers sounded "regimental call" and "assembly." From all sides parts of the regiment worked their way toward the area west of Bleid. One after the other the different companies came in. There were many gaps in their ranks. In its first fight the regiment had lost twenty-five percent of its officers and fifteen per cent of its men in dead, wounded, and missing. I was deeply grieved to learn that two of my best friends had been killed. As soon as the formations had been reordered, the battalions set off toward Gomery through the south part of Bleid.

Bleid presented a terrible sight. Among the smoking ruins lay dead soldiers, civilians, and animals. The troops were told that the opponents of the German Fifth Army had been defeated all along the line and were in retreat; yet in achieving our first victory, our success was considerably tempered by grief over the loss of our comrades. We marched south, but our progress was frequently halted, for in the distance we saw enemy columns on the march. Batteries of the 49th Artillery Regiment trotted ahead and went into position on the right of the highway. By the time we heard their first shots, the enemy columns had disappeared into the distance.

Night fell. Nearly dead from fatigue, we finally reached the village of Ruette, which was already more than filled with our own troops. We bivouacked in the open. No straw could be found, and our men were much too tired to search for it. The damp, cold ground kept us from getting a refreshing sleep. Toward morning it grew chilly—all of us were pitifully cold. During the early morning hours, my complaining stomach made me restless. Finally day dawned. Again thick fog lay over the fields.

Observations: It is difficult to maintain contact in fog. During the battle in the fog at Bleid, contact was lost soon after meeting the enemy, and it was not possible to reestablish it. Advances through fog by means of a compass must be practiced, since smoke will frequently be employed. In a meeting engagement in the fog, the side capable of developing a maximum fire power on contact will get the upper hand; therefore keep the machine guns ready for action at all times during an advance.

Fights in inhabited places often take place at extremely short ranges (a few yards). Hand grenades and machine pistols are essential. Provide fire protection before attacking by means of machine guns, mortars and assault guns. An attack in a village is usually accompanied by heavy casualties and should be avoided whenever possible. Pin the enemy down to the village by means of fire, or blind him with smoke and hit him outside the village or town.

Tall grain offers good concealment, but shining articles such as bayonets and cooking utensils may betray the location of troops. French security measures at Bleid were totally inadequate. Likewise, they failed to observe proper security precautions during this retreat and during the combat in the fields. After the first exchange, the German rifleman became imbued with a feeling of superiority *vis-à-vis* his French counterpart.

V: On the Meuse

Battles at Mont and in the Doulcon Woods

Following the battle of Longwy, the enemy was pursued in a southwesterly, then in a westerly direction. In the Chier and Othain sectors, we had short but violent fights. During these battles, the French artillery covered the infantry withdrawals by means of highly-concentrated and well-placed fire even at the sacrifice of the batteries themselves. During the night of August 28-29 the 7th Company of the 124th Infantry Regiment was on outpost duty south of Jametz. All outposts and pickets dug in. On August 29 the advance continued to the Meuse. During a rest, the 13th Engineers who were up ahead in the column, were attacked west of Jametz by strong enemy forces issuing from nearby woods. Violent hand-to-hand fighting ensued, the engineers attacking the enemy with shovels and axes. Heavy casualties were suffered on both sides. The 123d Grenadier Regiment and the 3d Battalion, 124th Infantry, also took part in the fight. The battle ended with the capture of the governor of the Montmédy fortress and two hundred men of the garrison who tried to cut their way through to Verdun. We passed by Montmédy.

East of Murveaux from positions on the west bank of the Meuse the French greeted us with shrapnel but inflicted little damage. Their fuzes were set too high. Towards noon we were on our way in the burning sun to Dun on the Meuse. The fire of the French artillery was growing stronger. The battalion deployed in the woods one mile east of Dun. Companies formed into columns among the tall trees. Shortly thereafter the French laid heavy artillery fire on this part of the woods. We could plainly hear the guns fire in the distance; then came the sound of the approaching shells which, a few seconds later, flew through the leafy cover over our heads and burst with a terrific roar, some against the trees, others deep in the ground. Fragments screamed through the air and chunks of sod and branches dropped on us. Now they fell very close, now farther away. At each detonation we huddled together and dropped onto the ground. The continual danger was wearing on us. The battalion remained where it was until evening. Our casualties were astonishingly few.

Ahead of us on the edge of the woods half a mile south of Dun, the 4th

Battery of the 49th Field Artillery Regiment, with which I had served a month before, was engaged in violent action against the enemy from a half-concealed position. It could not stand up to the French artillery which was superior in matériel, and the battery was suffering losses in men and equipment.

At twilight the 2d Battalion moved back to Murveaux. We spent the night in the open. My stomach was grumbling, for I had eaten nothing all day except for a handful of grain. We were short of bread.

On the morning of August 30 French artillery broke up our divine services. The artillery duel on the Meuse was growing in intensity. To our great joy, tractor-drawn 210mm batteries moved up and went into position; and soon their heavy shells were on their way to the enemy.

We spent the night of August 30-31 in cramped quarters in Murveaux. In the morning the 2d Battalion proceeded toward Sassey by way of Milly; crossed the Meuse over a ponton bridge; and, as advance guard for the 53d Brigade, took up the march to Mont-devant-Sassey. Soon after our arrival at that place a search of all the cellars netted us twenty-six French infantrymen. They belonged to the 124th Infantry Regiment—our regiment's number.

At the southwest entrance of Mont our advance infantry ran into heavy fire from the woods on the commanding heights west of Mont, Shortly afterward our own artillery began to fire toward Mont from the hill southwest of Sassey and caused casualties in our units. They were firing on the report of a mounted reconnaissance detachment which a half hour previously had been fired on from Mont. Some time elapsed before the mistake could be corrected.

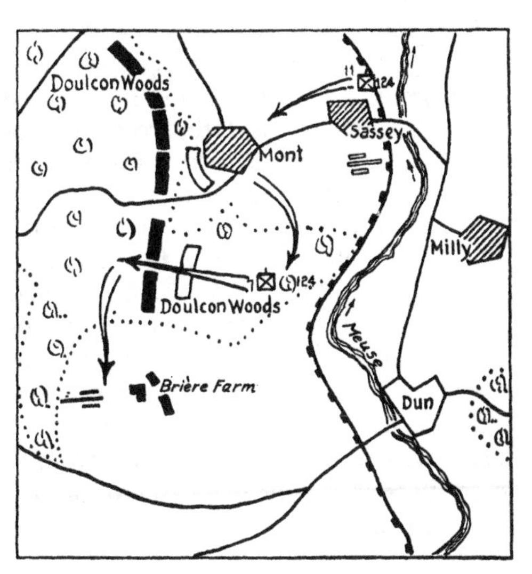

Sketch 3: The battle for Mont and Doulcon Woods.

One platoon of the 7th Company advanced to attack the enemy on the hills west of Mont, but heavy enemy fire stopped the attack in its tracks. Committing an additional platoon brought no better results. Established in their dominating position, the far superior forces of the enemy caused severe losses among the infantry climbing the steep slopes, especially since the assault elements were unable to return the fire.

After our attacks were repulsed, the 7th Company was withdrawn and ordered to the assistance of the hard-pressed 127th Infantry Regiment in the Doulcon woods a mile and a quarter south of Mont. The company moved through the village of Mont to the southeast, and moved behind a hedge in open column of file. Thus hidden from enemy observation, they climbed Hill 297. Scarcely had the company reached the Mont woods and closed up when French shrapnel forced it to hit the dirt. We found shelter behind trees and in depressions in the ground. There were no signs of the 127th Infantry.

At the order of the company commander, I moved with two men toward the southern edge of Doulcon woods in order to establish contact with the regiment. We were fired on several times before reaching the south edge of the woods, and we found no sign of our own people. Down below in the Meuse valley, Dun was under violent French artillery fire. Using the reports of the guns as a guide, we estimated the French artillery to be located behind a line of hills on the west bank of the Meuse. Neither our own nor the enemy infantry was to be seen at this time.

After my return the company advanced toward the west over a forest road. Upon reaching a clearing about a hundred yards wide, we established security elements in all directions and rested while retaining our march formation. Next the company commander sent out scouting detachments in various directions in order to find the whereabouts of the 127th Infantry Regiment. These were hardly out of sight (the company had been resting about five minutes) when the whole clearing was subjected to intense shrapnel fire. The shells rained down like a sudden thunderstorm. We tried to find shelter behind trees and used our packs to form improvised breastworks. The intensity of the bombardment made it impossible to move in any direction. Although the bombardment lasted several minutes, there were no casualties. Our packs intercepted a few of the missiles, and the bayonet tassel of one of the men was torn in shreds. It was a mystery to us how the French artillery so quickly learned of our location in the middle of the forest and how it was possible for them to lay their fire on us in such a short time. Was it just an accident?

At this moment one of my men from the scouting detachments returned with a badly wounded man from the 127th Infantry. The latter said that his regiment had retired hours ago and that, excepting dead and wounded, nothing remained in the woods up ahead. Two hours before several French battalions had marched past him going in a northerly direction and he believed that these troops must still be in the woods.

Under these circumstances our lone company's prospects were none too bright. Should we also return? The appearance of an infantry battalion on the road behind us solved our problem, and, after a conference with the battalion commander, we moved out in a westerly direction as battalion advance guard with my platoon as the point.

Five minutes later we heard heavy bursts of small-arms fire accompanied by considerable shouting. The sounds came from our right, and I estimated the distance to be about two-thirds of a mile. We turned toward the sound of the guns and moved over a narrow trail, both sides of which were covered with a heavy underbrush. On a straight section of trail we could make out some black objects a hundred yards ahead. Bullets whining about our ears answered the question as to their identity. We took cover in the brush, and the company deployed on both sides of the trail. The enemy maintained a heavy volume of fire but most of it was wild though we did suffer some injuries from ricochets. We advanced on our bellies through the thick underbrush and withheld fire until we were about 150 yards from the enemy position. Because of the dense underbrush I could see but few of my men, let alone control them. An increase in light indicated that we were nearing a clearing. Judging from the sounds ahead, we were about one hundred yards from the enemy. I charged ahead with my platoon and reached the clearing, which turned out to be so overgrown with blackberry bushes that we were unable to cross it. Violent enemy rifle fire drove us to earth, and we joined the fire fight with the enemy on the other side of the clearing. In spite of the close range, our targets remained hidden by the dense foliage and undergrowth. The two remaining platoons came up and we extended our skirmish line with two or three paces interval between men. The company commander ordered: "Continue firing and dig in." I noticed that the company commander, Lieutenant Bammert, was in the front line lying alongside a big oak. Any movement was out of the question, but fortunately for us the enemy was still shooting high. Even so, men were getting hit.

While some riflemen maintained a slow protective fire the remainder dug in. The condition of the soil did not make this an easy job, and branches and leaves kept raining down. Suddenly we heard firing from a new direction and from behind us. Bullets struck around me, throwing dirt in my face. The man on my left suddenly cried out and rolled on the ground in pain. He had been shot clean through. Crazed with pain he cried out: "Help! Aid men! I am bleeding to death!" I crawled to the wounded man, but he was past help. His face was distorted with pain and his hands clawed the ground until a shudder shook him from head to toe; thus we had lost another brave soldier. Since we were in poorly covered positions, this fire from both front and rear tended to unnerve us. It appeared that elements of our battalion had started their fire-fight as soon as they came within range of the enemy, and the thick underbrush made it difficult to correct this error. On the right the sound of battle increased, and this brought an increased volume of enemy fire. A bullet smacked into the blade of the shovel with which I was digging. A few moments later First Lieutenant Bammert was hit in the leg, and I took command of the company. An attack was being made by German forces on our right; for we could hear drums, bugles, shouting, and the methodical fire of the French machine guns. It was a welcome respite. I ordered the 7th Company to attack, passing around the left of the clearing. The troops rushed forward, glad to get out of a bad spot and determined to fight it out. The enemy decided to defer our meeting and threw a few shots at us; but by the time we had reached the far side of the clearing, he had vanished into the underbrush. We started off in pursuit, my immediate objective being the southern edge of Doulcon woods because from there we might be able to inflict additional damage on the retreating enemy as he crossed the open terrain. Thinking that the whole company was right behind me, I hurried on as fast as possible with the leading squads but failed to catch up with the enemy before we reached the southern edge of Doulcon woods. Ahead of us, to the south on the next rise and on the far side of a wide meadow, was Brière farm. Behind the same rise and to the right of us we saw a French battery firing up the Meuse valley in the direction of Dun. Strange to say, no enemy infantry could be located. Judging from appearances, they had retired into the woods on the west. We now had lost contact within the company, and my total available force was twelve men. From the left a scouting detachment of the 127th Infantry came up and informed me that the 127th Infantry was about to attack from the woods in the direction of the Brière farm. Soon we saw skirmish lines advancing on the left. My problem was whether to wait for the remainder of the company or to attack the battery ahead with my twelve men. I decided on the latter in the hope that the remainder of the outfit would follow. By dashes we reached a depression; and some seven hundred yards west of Brière farm we started climbing in the direction of the French battery. Judging from the sound of the guns, we were separated by a scant hundred yards. On our left the advance elements of the 127th Infantry were closing in on the farm. It was getting dark. Suddenly our own troops opened fire on us from the farm; the 127th must have taken us for Frenchmen.

The fire became heavier and forced us to the ground. We tried to set them right by waving our helmets and handkerchiefs, but it was no use.

There was no cover close to us, and the rifle bullets were striking close about us in the grass. We pressed our bodies to the ground and resigned ourselves to being fired on by our own people—for the second time in the course of a few hours. Seconds seemed like eternity, and I could hear my men groan as the bullets whistled over us. We prayed for darkness since its cover offered us our only chance of salvation. Finally they ceased firing. In order not to draw any more fire, we stayed where we were, and then after waiting several minutes we crawled back into the hollow to our rear. We made it! My twelve men were unscathed.

It was now too late to attack the French battery and I had lost my stomach for it. The moon shed a dim light through sparse clouds as we headed back into the Doulcon woods, the scene of the afternoon's battle. We found no trace of the company. Later, I learned that a soldier had told the first sergeant I had been killed in the fight in the woods. The sergeant had assembled the outfit and marched it back to our battalion in the vicinity of Mont.

Passing through Doulcon woods, we heard the moans of wounded men all around us. It was a gruesome sound. A low voice from a nearby bush called *"Kamerad, Kamerad!"* A youngster from the 127th lay with a breast wound on the cold stony ground. The poor lad sobbed as we stooped over him—he did not want to die. We wrapped him in his coat and shelter half, gave him some water, and made him as comfortable as possible. We heard the voices of wounded men on all sides now. One called in a heart-breaking way for his mother. Another prayed. Others were crying with pain and mingled with these voices we heard the sound of French: *"Des blessés, camarades!"* It was terrible to listen to suffering and dying men. We helped friend and foe without distinction, gave them our last piece of bread and our last drop of water. We were unable to move severely wounded men out of this rough terrain without stretchers, for our makeshift carry methods would only result in an added number of painful deaths. Exhausted and hungry we reached Mont shortly before midnight. The village had suffered heavily, with several houses totally demolished by the bombardment. Dead horses were lying in the narrow streets. In one of the houses I ran across a medical company. I described the location of the wounded in Doulcon woods for their commander and made arrangements for their care. One of my men volunteered as guide. Then I looked about for shelter for the night. There was no trace of my own battalion.

In one of the houses we saw a light shining through the closed shutters. We went inside, and found a dozen women and girls who seemed frightened at our appearance. In French I asked for food and a place to sleep for myself and my men. Both were provided, and soon we were sound asleep on clean mattresses. At daybreak we started looking for the 2d Battalion and found it just east of Mont.

There was general amazement at our return, as we had been given up for lost. First Lieutenant Eichholz now took command of the 7th Company. In the evening we found quarters in Mont, and our company placed sentries at the southwest entrance. I slept in a bed in regal fashion, but not until I had forced the local French billeting agent to disgorge a couple of bottles of wine for Hänle and myself. Bedbug bites were the souvenirs of my princely couch.

Observations: The attack on the engineer company resting at the head of the main body of troops teaches us that all units of a group must provide for their own security. This is especially true in close terrain and when faced with a highly mobile enemy.

In the woods east of Dun the 7th Company was under heavy French artillery fire for a considerable period. Had one of the shells hit the column, at least two squads would have been annihilated simultaneously. With the increased power of modern weapons, increased dispersion and digging of foxholes is vital to the safety of any unit. Begin digging-in before the first enemy bombardment. Too much spade work is better than too little. Sweat saves blood.

As is shown by the example at Mont a thorough search of any locality previously occupied by the enemy is necessary. The twenty-six captured Frenchmen were perhaps shirkers, or they may have been left behind to ambush us as we moved through the town.

The report of a cavalry reconnaissance detachment that it had been fired on from Mont a half hour before caused our own artillery to fire into Mont while it was actually in the possession of the 124th Infantry. Needless losses resulted. Artillery-infantry liason must be maintained. The artillery must maintain uninterrupted observation over the field of battle.

To march or halt in a closed column within artillery range of the enemy is bad practice, as shown by the French bombardment of the company halted in Doulcon woods. With modern artillery very heavy casualties would have resulted. The fight in Doulcon woods emphasizes the difficulties of forest fighting. One sees nothing of the enemy. The bullets strike with a loud crash against trees and branches, innumerable ricochets fill the air, and it is hard to tell the direction of the enemy fire. It is difficult to maintain direction and contact in the front line; the commander can control only the men closest to him, permitting the remaining troops to get out of hand. Digging shelters in a woods is difficult because of roots. The position of the front line becomes untenable when

—as in the Doulcon woods—one's own troops open fire from the rear, for the front line is caught between two lines of fire. In advances, as well as in forest fighting, it is advisable to have a maximum number of machine guns well up forward. It will be necessary to fire the machine gun while on the move in case of chance encounters or while engaged in the assault.

Chapter 2

BATTLES AT GESNES, DEFUY WOODS AND REMBERCOURT

I: THE FIGHT AT GESNES

In the earliest hours of September 2, 1914, the battalion headed for Villers-devant-Dun, where we got a short rest. Then the battalion hurriedly rejoined the regiment and, under a hot sun, marched through Andeville and Remonville to Landres. The enemy had retreated and the Meuse lay behind us. Morale was high in spite of the battles and exertions of the last few days. The band played as if we were on practice maneuvers. To the south in the direction of Verdun we could see artillery flashes and hear the shell bursts. We marched west in the heat and dust.

At Landres, in the afternoon, we suddenly turned southeast. Over miserable paths and through heavily wooded terrain, the 124th Infantry hurried to the assistance of the hard-pressed 11th Reserve Division. In the woods one mile northwest of Gesnes, French artillery got the range and greeted us with a shower of shrapnel. The battalion halted and I was sent ahead in the direction of Gesnes to find a road that offered some cover from this artillery fire. Accompanied by a sergeant, I moved through a dense brush to the south edge of the woods where we were forced to take cover because of heavy fire raking the edge of the woods from the right. We continued to the left and discovered a fairly well protected road. On returning we found the battalion had moved. Hänle was waiting alone with the horse and reported that the column had marched off to the right. Ahead enemy shells were striking along the edge of the woods. Accompanied by Hänle and the sergeant, I rode forward toward Gesnes, in order to overtake the unit, on the road I had just reconnoitered. On leaving the edge of the woods I could not locate the battalion. Perhaps it had already passed over the hill on its way to Gesnes. An officerless company of the 11th Reserve Division asked me to assume command. Soon three additional officerless companies were following me. Deploying, I led my fairly large force from the edge of the woods in the direction of Gesnes. On a slope three quarters of a mile northwest of Gesnes we halted and reorganized and the result was quite imposing. The ridge of the hill ahead was under intense fire from French rifles, machine guns, and artillery. Our own troops appeared to be engaged there. While my new formation was being reorganized, I rode forward and tied my horse to a bush on the protected slope close behind our own skirmish line. Up ahead I found elements of the 1st Battalion, 124th

Infantry, mixed with troops of the 123d Grenadier Regiment, all engaged in violent fire fight with the enemy on the hills south and southwest of Gesnes. Our own attack had stalled in the face of heavy small arms and artillery fire and the men were digging in.

The enemy was well concealed and hard to locate even with field glasses and his artillery made life miserable for us. No one had seen anything of the 2d Battalion. Was it still in the woods to our rear? I galloped back. On the way I met the colonel of the 123d Grenadier Regiment and reported to him concerning the situation on the hill, giving him the location of the battalion that had placed itself under my orders. To my intense regret, an older officer was given command of this outfit and I was left free to continue my search for the 2d Battalion, 124th Infantry. I could not find it and rode back to the front line on the hill thirteen hundred yards northwest of Gesnes, collecting the parts of the 1st Battalion, 124th Infantry, that remained there. Soon I had about a hundred men with me.

The French batteries opened with rapid fire, and during the next few minutes hell broke loose around us; then one after the other, the enemy batteries ceased firing and finally all were silent. Night fell and except for sporadic flare-ups the rifle fire died away. I continued looking for the 2d Battalion on the hills west of Gesnes until late, but luck was against me and I returned to my men. All were worn out and hungry for they had not eaten since early morning. Unfortunately I could not supply them with rations and I had my doubts as to whether the kitchens had managed to get through Gesnes Woods. My intention was to start back at daybreak in the direction of Exermont where I hoped to locate my regiment. The night passed without incident and toward morning we had a decided drop in the temperature. My complaining stomach served in place of an alarm clock.

At dawn, French small-arms fire started in again over a broad front. We retired in the direction of Exermont, and in a hollow a mile and a half east of that place I found the regimental CP near which I located the 2d Battalion, 124th Infantry, which was in regimental reserve. After reporting I was given a new job. The battalion adjutant was a casualty, and I was ordered to take his place. The food situation was no better here than up forward and I ate some wheat grains to quiet my protesting stomach.

Infantry small-arms fire could be heard again, but the artillery had ceased firing. About 0900 the battalion commander took me with him on reconnaissance. The 1st and 2d Battalions held the ridge between Exermont and Gesnes. During our ride we had ample opportunity to see the results of the last day's fighting. Dead were everywhere, and among them we recognized the bodies of Captain Reinhardt and Lieutenant Holmann who fell on the preceding day. Our own front line was now dug in and there was little to see of the enemy who still held Tronsol farm. We returned to the battalion.

My next job was to find the battalion kitchens and bring them up. This was an imperative task, for the troops had not eaten for more than thirty hours. To complicate matters, no one knew where the field kitchens were. I began by searching Gesnes and Romagne Woods and then proceeded to Romagne. The latter was full of vehicles belonging to the 11th Reserve Division. My next stop was at Gesnes because I remembered that the kitchens had been ordered there via Exermont, and I had a feeling that I would find them around our front lines. Gesnes was empty, and I headed toward Exermont which lay in the valley between the two fronts. Firing from the heights on both sides had ceased, and one mile southwest of Gesnes I ran into the entire 2d Battalion combat train. My hunch was correct for they were ahead of the front line. Shortly afterward some scouts arrived with the information that the regiment was moving forward in a quarter of an hour. Under these circumstances, I left the kitchens where they were.

The hills around Tronsol farm were taken without further opposition. The enemy had withdrawn by the south leaving a few dead and wounded behind. The regiment bivouacked under canvas in the vicinity of the farm. My horse got a stall in the stable. He was in need of care after several strenuous days and cold nights.

II: Pursuit Through the Argonne; Fight at Pretz

On September 4 we marched to Boureuilles by way of Eglisfontaine—Very—Cheppy and Varennes. The roads bore testimony to a hurried enemy retreat—abandoned rifles, packs, and vehicles. Because of the heat and the dusty condition of the road, we made slow progress, reaching Boureuilles late at night. During the night my ailing stomach again robbed me of sleep. The next day we marched through the Argonne to Briceaux, passing Clérmont and Les Ilettes. We did not make contact with the enemy, and we knew that their rear guard had pulled out an hour ahead of us. Verdun was seventeen miles northeast of Briceaux. We were well quartered in Briceaux, but nobody was very hard to please. A mattress and a bite to eat were adequate. Captain Ullerich took command of the 2d Battalion. At daybreak, September 6, we sent out a mounted reconnaissance detachment which was fired on from the woods a bit south of Briceaux. At about 0900 the regiment left Briceaux, moving deployed to the southwest. At Longues Bois our leading elements ran into the enemy. The 1st Battalion attacked and quickly seized the

Triancourt—Pretz highway. A few French troops were taken prisoner.

The 1st Battalion pushed down the road toward Pretz with the 2d Battalion following. Tall woods lined both sides of the highway. To the left there was violent fighting. On reaching the southern edge of the woods, the 1st Battalion ran into stronger enemy forces. A hot fire fight started at a couple of hundred yards' range. But once again the French artillery made the going rough. Apparently they were well supplied with ammunition and they delivered a most effective and flexible fire. The 2d Battalion entered the woods, but the artillery soon made that place untenable.

Toward noon the 2d Battalion was ordered to advance along the southwest edge of the woods to a point a mile and a quarter west of Pretz and then to attack on the right of the 1st Battalion and take Hill 260 (860).

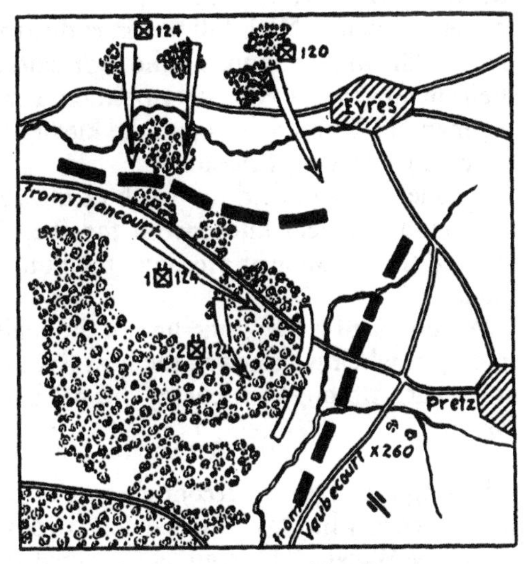

Sketch 4: The attack on Pretz.

We moved out with Lieutenant Kirn in command of the point. We reached Hill 241 (795) without meeting the enemy. From here we had to ride through tall bushes which almost covered the narrow path. About one hundred yards from the edge of the woods, we suddenly saw a strong French reconnaissance detachment ahead of us. Both sides opened fire at close range, and the French retired without inflicting any casualties on us. On looking around we found that we had lost contact with the battalion. In order to reestablish it, the point halted and I rode back and found the battalion lying off to the left of the woods. I reported the latest action and the retreat of the enemy. The march on Hill 241 (795) continued but after progressing a few hundred yards, French artillery fire drove the battalion to the ground. For several minutes we were subjected to a veritable hail of artillery fire with all movement out of the question. The men took cover as best they could behind trees, in hollows, and even behind piled packs. We suffered some casualties.

When the fire became less intense, I galloped through the woods on the left to reestablish contact with the 1st Battalion. The woods proved too swampy. Failing to accomplish my mission, I returned and worked my way forward on foot along the eastern edge of the woods. I was frequently fired upon by the enemy, who had occupied the rise 350 yards east of the woods. At last I located the 3d Company which had held up its attack pending our assault.

Immediately on my return the battalion launched its attack in the direction of Hill 260 (860) with the 6th and 8th Companies in assault. The French abandoned their positions and fell back. Even the French artillery which had been the bane of our existence all day was no longer much in evidence. We seized Hill 260 (860) and poured fire on the retreating enemy. Nightfall put an end to the fighting. Scouts were sent out and the units dug in. To the right and ahead of us we saw piles of shells in an abandoned battery position. I was sent back to report to the regimental CP and to bring up the kitchens. The men had had nothing to eat since leaving Briceaux.

Colonel Haas praised the work of the 2d Battalion.

I found the field kitchens on the Pretz-Triancourt highway. They reached the battalion at 2100, and the hungry men finally got some hot food.

We now had a telephone line to the regimental CP and it was after midnight when we received the next day's order. Our own scouts came and went. Although the enemy did not bother us, there was little time for rest.

III: Attack on Defuy Woods

During the night our reconnaissance elements were able to determine that the enemy had taken up a defensive position some two miles distant in Defuy Woods. The regiment ordered the 2d Battalion to cross the highway at 0600 and take the woods. Units from the 123d Grenadier Regiment were to advance on our right.

At H hour the battalion attacked with two companies (6th and 7th) in the assault and two companies (5th and 8th) echeloned to the left rear. Our left flank advanced toward the northeast corner of the woods. I rode between the 6th and 7th Companies. There was no sign of the grenadiers on our right. At this point we received this order: "Stay the advance. Remain where you are."

I transmitted the order and galloped back to the regimental CP on Hill 260 (860) to find out the why and wherefore of the order. Colonel Haas wanted the attack held up until the 123d could get going and he had no idea when that would be. In the meantime the French artillery had become active and was laying its fire on the two reserve companies which were bunched together in the open. The French artillery observers had an excellent view of our lines from the northern edge of the woods.

I dashed forward with the battalion order for the assault echelon to entrench in the potato fields and vegetable plots. On my way back, a French battery drew a bead on me, and I had to zigzag to avoid getting hit by the shrapnel it was throwing my way.

The French artillery fire, with medium guns adding its weight, increased in intensity. The 5th Company was lying on the ground in closed column, and a single shell wiped out two entire squads. The front-line units were well concealed and dug in and so did not share the fate of the 5th Company.

A battery of the 49th Artillery Regiment, which took up the fight from positions near Hill 260 (860), received a bad mauling from the French counterbattery fire.

The battalion and regimental command posts were located close together in a cut on the highway a mile and a quarter northeast of Vaubencourt. It did not take the French batteries long to lay an extremely heavy concentration on the cut. And no wonder! The heavy traffic in messengers and horsemen, not to mention the numerous OPs, had given the location away. This harassing fire kept up for hours, and we called off the attack. My bunk for the night was the ditch running alongside the road where I tried to catch up on some lost sleep. In spite of my frayed nerves, I was no longer bothered by the noise of the enemy artillery. The fire had torn up most of the trees in our vicinity, but our casualties were small. Just before dark an order to press the attack put an end to our idle brooding. With the 123d Grenadier Regiment on its right, the 3d Battalion moved up on the right of the 2d Battalion. While these maneuvers were going on the French artillery fire slackened and then died away.

I went ahead and set the battalion in motion. Strange to say, we encountered no fire from French small-arms or artillery. Had they cleared out again?

The front line—a skirmish line at four-pace intervals—crossed the low ground six hundred yards northwest of the woods and climbed the slope. On the right the grenadiers and the 3d Battalion were abreast of each other. The reserve (1st Battalion 124th Infantry and Machine-Gun Company) followed a couple of hundred yards behind the attacking troops.

I rode behind the 7th Company which was on the extreme left. Dusk was falling.

All remained quiet until we were some 150 yards from the woods, when, to our surprise, the French opened on us. A brisk fire fight resulted, and the company reserves were rushed up and hit the dirt alongside the men in the front line. Nobody had much protection, and the heavy fire drove the entire regiment to cover. Some of the machine-gun platoons unloaded their weapons and opened up on the French. They were trying to shoot over the heads of our own front line, but the yells that came from up ahead told us that our machine guns were firing into our forward elements. The entire action was over in less time than it takes to tell it.

I was mounted on the extreme left flank of the battalion. From there I galloped across to the machine guns, had them cease fire, dismounted and handed my horse to the first man nearby and then took charge of a platoon which I led to the left of the battalion. There the machine gun was promptly emplaced and opened on the enemy. Aided by this additional fire support we pressed the attack together with the units on the right. Now all signs of fatigue and exhaustion had left us; our fighting pitch was at fever heat and we wanted to come to grips with the enemy. Rifle fire tore gaps in our ranks, but did not stop us, and we smashed into the woods only to find that the enemy had again broken off and abandoned his positions. The regiment ordered the woods cleared but the thick undergrowth did not make that an easy task. Why not go around the woods and cut the French off? It did not take me long to decide. With two squads and the heavy machine-gun platoon I climbed the slope on the left of the woods. Here there was no underbrush to slow us down and we could not conceive of the enemy moving through the woods as quickly as we were advancing around them. Out of breath, we finally reached the eastern corner of the woods. It was still light enough to shoot and we observed that we had a field of fire that covered the southern exit of the woods for a distance of several hundred yards. Feverishly we got the heavy machine guns into position and the riflemen concealed themselves close to the eastern corner along the edge of the woods. The enemy was expected to emerge from the woods at any moment. To the right and rear we could hear German signals.

Minutes passed with no signs of the enemy and slowly the light faded. Over on the left the blazing buildings in Rembercourt lit the sky. My conscience was bothering me for I had taken the heavy machine-gun platoon without the regimental commander's permission. Since all prospects of a fight had vanished, I released the platoon and returned it to its company. It had scarcely left when one of the riflemen pointed out a column of men, visible in the light of the fires from Rembercourt, passing over the bare crest of the hill about one hundred and sixty yards away. Frenchmen! I could distinguish their képis and bayonets with my field glasses. There was no doubt that the enemy was withdrawing in close order. I regretted sending the machine-gun platoon away only a few minutes before but it was too late to recall the order.

My sixteen rifles opened with rapid fire at the enemy. Contrary to our expectations, the French did not break and run but rushed at us yelling *"en avant!"* Judging from the volume of sound, there must have been one or two companies of them. We fired as fast as we were able but they kept coming. I dragged back some of my men who were about to retire on their own hook. Apparently our fire was forcing the enemy to hit the dirt. It was hard to pick out the enemy soldiers on the level meadow land and in the light of the burning buildings in Rembercourt. His leading elements were some thirty or forty yards in front of us. I had made up my mind not to give way to their superior numbers until they were ready to deliver a bayonet charge. The charge never materialized.

Our fire had dampened the enemy's enthusiasm for an attack. The battle cry of *"en avant!"* ceased. Only five French machine-gun pack horses, carrying two machine guns continued to advance as far as the edge of the woods where they were captured. It grew quiet around us. Apparently the enemy was retreating toward Rembercourt. A reconnaissance detail which went out and gathered in a dozen prisoners reported that some thirty dead and wounded French littered the field.

Where was the 2d Battalion? Apparently it had not pushed on through Defuy Woods as per order. To reestablish contact I went back to the northeast corner of the woods with two men who brought the prisoners and pack horses along. I left the remainder of my two squads in position.

On the way, I ran into the regimental commander. Colonel Haas was not at all pleased with what had happened on the edge of the woods. His opinion was that I had not been firing on Frenchmen but on elements of the Grenadier Regiment. Even the prisoners and pack horses with their machine guns failed to convince him.

Observations: The attack on September 7 against the Defuy Woods had to be carried out over terrain that was two miles in width and offered little or no cover. On regimental order the attack was held up because the unit on the right had not advanced. At the same time, the French artillery began a heavy bombardment. The deployed elements of the 2d Battalion quickly took cover in the potato field and found protection from this fire by digging with their spades. They had no casualties in spite of the heavy day-long artillery bombardment. On the other hand, the close order of a reserve company led to heavy casualties from enemy artillery fire. This teaches us again that no massing together is permissible within enemy artillery range and reemphasizes the importance of the spade.

The regimental and battalion command posts were situated close together in a highway cut. Their location was betrayed to the enemy by the large number of messengers who converged there from all directions. The enemy reacted properly and plastered the place with artillery fire. Command posts *must* be dispersed.

All traffic, either on foot or mounted, must approach by roads and paths hidden from enemy observation. The enemy must not be able to distinguish a CP; hence do not choose a conspicuous hill for its location. After dark French artillery fire ceased. It was displacing to the rear, probably to avoid capture in event of a German night attack. The French infantry let the Germans approach to within 150 yards before opening up, engaged in fire fight of a few minutes' duration and then, sheltered by the woods and approaching darkness, broke off contact and withdrew. Our losses were heavy; at the end of the day on September 7 we had 5 officers and 240 men on our regimental casualty list.

In the excitement of the fight, elements of the machine-gun company fired over the heads of the crowded infantry line lying four hundred yards ahead of them on a rising slope in an effort to reach the enemy on the edge of the woods six hundred yards away. This created a very dangerous situation for the front-line units. Believing that enemy resistance was over, we abandoned our assault formation in depth and brought the reserves and machine-gun units up to the front line. We paid a heavy price for this tactical error when the enemy opened on us with well-aimed rifle fire at a range of one hundred fifty yards.

In similar situations some of the soldiers will often lose their nerve and break for cover. The commander must take vigorous action, using his personal weapons if necessary.

IV: Battle At the Defuy Woods

The regiment ordered the 3d Battalion to establish a defensive position along the south edge of Defuy Woods with the left wing resting on the eastern corner of the woods. Astride the woods, the 2d Battalion was to prolong the line on the left of the 3d Battalion. The 1st Battalion established the regimental reserve line north of the Defuy Woods. The regimental CP was on the left of the 1st Battalion.

The sector assigned to the 2d Battalion—a long, barren ridge devoid of cover—did not please us at all. Positions on this ridge would be especially exposed to French artillery fire. We would have preferred the 3d Battalion's position in the woods.

Our recent experiences indicated but one way of keeping casualties down—the deep trench. Company sectors were assigned, and the company commanders—three of them young lieutenants—were impressed with the vital necessity of seeing that the men dug without regard to fatigue. The main part of the work had to be finished before midnight. Between then and daybreak a few hours' rest could be taken, but shortly before dawn the work had to be continued. The trenches were to be five and a half feet deep.

Soon the whole battalion was hard at it. The violence of the enemy artillery fire the day before had impressed us all with the value of spade work. Even the battalion staff consisting of battalion commander, adjutant, and four messengers dug itself a twenty-foot trench back of the center of the 8th Company which was located in front and on the right. The work was tiring, the ground proved to be hard as a rock, and it was almost impossible to accomplish anything with the short spades. Since only a few picks were available our progress was very slow. The men had not eaten since 0500, and at 2230 the battalion commander sent me to Pretz to bring up the kitchens. I got back at midnight with mail as well as food. This was the first mail we had received since the outbreak of the war.

At the end of several hours' digging the trench was some eighteen inches deep, which was certainly insufficient protection from hostile artillery fire. That meant more work before morning; but right now, at midnight, the men were completely exhausted. First they must be fed and get some rest. The kitchens arrived, the men were given food, and mail was distributed. In the narrow trench and by candlelight the men read letters which had been mailed from home weeks before. The letters came from another world, and yet we had not been away for years—just a few eventful weeks. The meal finished, back we went to our picks and shovels. The battalion staff did not rest until early morning, at which time our trench was about forty inches deep. Our blistered hands ached, and we were so tired that we felt no discomfort in dropping off to sleep on the hard ground in the quiet of an early September morning.

And now the companies were back at work! On the eastern edge of the woods we saw a gun section of the 49th FA going into position on the boundary of the 2d and 3d Battalions and some thirty yards behind the front line.

On September 8 there was no activity during the early morning hours. On the other side of the valley, the glasses showed enemy defense positions on Hills 267 and 297 (west and northwest of Rembercourt). We had visual contact with our neighbors on the left, the 120th Infantry Regiment located on Hill 285; and provisions had been made to sweep the six hundred-yard gap with fire. A heavy machine-gun platoon had gone into position in our sector. The 5th and 8th Companies were in the front line; the 6th and 7th Companies were echeloned behind the right and left flanks respectively. The battalion commander took me on his tour of inspection, and we found the men working hard. In some places the trench was now four and a half feet deep.

The French artillery opened up at 0600 and the heavy fire directed at us put all their previous efforts in the shade. The air was full of sound and fury, and the ground around us shook as if an earthquake had hit us. Most of the fire was with time fuzes which burst over us, but some impact fuzes were also being used. We lay huddled in our miserable trench with little protection against fragments from shells that burst in our vicinity. The bombardment maintained this intensity for several hours. On one occasion a shell struck the slope above us and rolled back into our trench, but luckily it was a dud. All of us were working to deepen the trench, and we used any tool that came to hand—picks, shovels, knives, mess tins, and even bare hands. One could see the men cringe as a shell burst in their vicinity. Along about noon, the enemy fire slackened and gave us our first opportunity to send runners out to the companies. Everything was in order, and there were no signs of French infantry. Fortunately, the losses were considerably less than our early estimate two to three per cent. The enemy soon stepped up the intensity of fire again. He must have had enormous quantities of artillery ammunition on hand. In contrast to this, our artillery was silent during most of the entire day, a silence imposed by a dearth of ammunition.

The French artillery kept it up all afternoon, but by now our trenches were seventy inches deep. Some men had dug fox-holes for themselves in the front wall of the trench. Not even time fuze fragments could reach them there, and with but twenty inches of hard soil above them they had protection even against shells with impact fuzes.

Toward evening the enemy raised his volume of fire to a terrific pitch and threw everything he had at us. A thick pall of black smoke from his medium artillery drifted across our positions. Shells ploughed up the slope and filled the air with dirt and stones. This might have been their preparation for an infantry attack. "Let them come," we said; we had been waiting for them all day.

The French artillery ceased fire as suddenly as it had begun, and the infantry attack did not come. We crawled out of our holes, and I made a tour of the four companies. The casualties were surprisingly small (sixteen men out of the battalion); and in spite of the severe nervous strain, the men were in the best of spirits. Their digging before and during the bombardment had paid big dividends.

The last rays of the setting sun lighted up the battlefield. Over to the right we saw the two guns of the 49th Field Artillery and about them their crews dead or badly wounded. The platoon's cover was such that shooting was impossible. It looked just as bad for the 3d Battalion in the woods on our right. There dense undergrowth made entrenching practically impossible. Concentrated French artillery fire, especially flanking fire, had had its effect heightened by damaged trees which fell on the troops and played havoc with various companies.

I went back to the regimental CP for orders and food. Colonel Haas was badly broken up over the heavy casualties in the 3d Battalion, which had to withdraw from the woods. The 2d Battalion was ordered to hold the hill to the east of Defuy Woods without support on either flank. In conclusion, Colonel Haas said, "The 124th Regiment will die in its positions."

On my return to the battalion the right flank of the 8th Company was ordered refused. The 6th Company took up a position with its front along the eastern edge of Defuy Woods and dug in. The other unit continued improving the position and the field kitchens arrived shortly before midnight. Again they brought mail with them. It was a repetition of the preceding night, the men resting for a few hours on the bare ground. The following day, the French artillery began firing at the same hour as on September 8 but from our good deep position we paid small attention to it. Part of the time we had telephone connections with the regiment, but shells kept breaking the line. I spent considerable time with the 5th Company and examined the enemy's positions with Sergeant Bentele, of the 7th Company. The French artillery was principally in open positions, and even the French infantry was exhibiting a great lack of caution. I prepared a report, complete with sketches, and sent it through the battalion to the regiment along with the request that artillery observers be sent forward to the 2d Battalion.

The left of the 120th Infantry was on the south slope of Hill 285 and some six hundred yards away, with the French across the way along the railroad tracks. French reserves were massed in a cut half a mile west of Vaux Marie station. From a knoll to our left we could reach them with flanking fire and probably cause considerable damage. I suggested this to the machine-gun platoon commander, but he had his doubts and declined. Taking matters into my own hands, I assumed command of the platoon, being well aware that we would have to work at top speed if we wished to avoid countermeasures from the French artillery. Within a few minutes our machine guns were firing into the massed enemy reserves, causing much confusion and a number of casualties. Our mission accomplished, we pulled out in a hurry and headed for cover; French counterbattery fire that followed hit empty positions. We suffered no losses, but during the operation the machine-gun platoon commander complained to the regimental commander regarding my arbitrary assumption of command. My explanation to the regimental CP was satisfactory and the matter ended.

During the day several artillery observers arrived in our sector. The layout of the French artillery was given them, but their ammunition supply was so small that the enemy artillery was not disturbed by our weak fire. However, one of our heavy batteries forced the enemy batteries located in Rembercourt to displace.

The evening was a repetition of the previous day. The French artillery "kissed us goodnight" by firing enormous quantities of ammunition. Then silence reigned. As far as we could make out, the French artillery was displacing to the rear.

We resumed our work in order to make our shelters bombproof, and several detachments went into the woods to cut trees. Fortunately, our casualties were lower than the day before, some occuring in the 6th Company as a result of flanking fire delivered against it. At about 2200 the field kitchens arrived and 1st Sergeant Rothenhaeussler of the 7th Company brought a bottle of red wine and a bundle of straw. Shortly before midnight I lay down on the straw close to the Battalion CP.

Observations: The 3d Battalion paid dearly for having established itself close to the southern edge of the woods. In this position it suffered extremely heavy casualties and had to be withdrawn during the night of the 8th. The heavy French artillery fire produced devastating results among the troops stationed in and on the edge of the woods. These units were not dug in properly. Many shells, which on a bare ridge would have been harmless "overs," caused damage among the troops by striking and exploding among the treetops. The forward edge of the woods was a death trap, and the French adjustment of fire there was very easy. Today artillery fuzes are even more sensitive and the losses in a similar case would be even greater.

In contrast to this, the 2d Battalion's pick-and-shovel work on the barren hill paid large dividends. In spite of an atrillery bombardment lasting for hours, our casualties were very small. Time fuzed shells were quite unpleasant, for many of their fragments flew straight into the trenches.

The hard soil in the 2d Battalion sector made digging difficult. It required all command powers, as well as personal example on the part of unit commanders, to force the tired and hungry men to dig their utmost during the night of September 7-8.

From September 7 to 9, the French artillery expended considerable quantities of ammunition. It had full stores to draw on, for the main sector dumps were close at hand. On the German side, we were running short of artillery ammunition and consequently the long arm could not render adequate support to the infantry.

Modern defense organization differs greatly from 1914. Then we had a front line with the remaining troops disposed in a second line. Today (1937) a battalion position consists of an outpost line and a main battle position through which the forces are organized in great depth. In an area eleven hundred to twenty-two hundred yards wide and deep, we have dozens of mutually supporting strongpoints garrisoned by riflemen, machine guns, mortars, and antitank weapons. These dispositions cause the enemy to divert his fire and the defense to concentrate its own fires. Local maneuver under covering fires is possible, and aggressive countermeasures can be instituted should the enemy succeed in penetrating the main battle position. The enemy has a long and very difficult road before breaking through.

V: Night Attack From September 9-10, 1914.

I fell asleep on my straw couch. Around midnight I awoke with a start. Combat raged ahead and to the left of us. It was raining cats and dogs and I was soaked to the skin. Off to the left I saw signal lights blinking and heard the continuous chatter of rifles. A messenger informed me that the battalion commander was at the regimental CP.

The sound of firing came closer and I began to wonder if the French were making a night attack. To find out what was going on, I took a runner with me and headed in the direction of the noise. Suddenly, fifty to sixty yards ahead, I saw human forms approaching us in a column of twos. I thought them to be French, who had penetrated the gap between the 124th and the 120th Infantry Regiments and were trying to strike the 2d Battalion in flank and rear. As they came closer and closer, I wondered what we should do. I decided to rush off to the right to inform Count von Rambaldi, Captain of the 6th Company, of the situation and request that a platoon be placed under my orders. The request was granted and I deployed my men and approached the enemy. When the light of distant flares allowed us to distinguish the outlines of the column, I ordered my men into position with rifles unlocked. I was still uncertain as to identities and challenged the column when it was some fifty yards distant. The 7th Company answered back. The company commander, a young lieutenant, was withdrawing from his position (echeloned to the left rear of the battalion) and was trying to move a quarter of a mile to the rear. His explanation was that there was going to be a fight and that his company was in the second line. Little pleased by his actions, I gave him a short tactical lecture. I still get the cold shivers when I think that I came so close to firing on my own recruits.

Shortly afterward the battalion commander returned from regimental headquarters with the regimental order for a night attack. Our battalion (in front line in the regiment) was to take Hill 287 (950), about five hundred yards north of Rembercourt, by storm. The adjoining regiments (the 123d Grenadier Regiment on the right, and the 120th Infantry Regiment on the left) were to attack at the same time. The time of the attack had not been decided on, but the battalion was to get ready immediately. The order promised a release from the hell of French artillery fire. The objective was not far away, and we wished that the French artillery positions on the hills around Rembercourt were also included.

In a pouring rain and in pitch darkness, the battalion got ready for the attack on the left of the former sector. Bayonets were fixed, rifles unlocked. The password was "Victory or Death." On the left there had been activity for quite some time. Rifle fire flared up and then died in one sector only to come to life somewhere else.

The 1st Battalion had come up. The regimental commander was with the 2d Battalion. Our information about the enemy was limited to a knowledge that he was along the railroad and in cuts south of the railroad and along the Sommaisne-Rembercourt road. Our men waited anxiously for H hour. By this time they had been soaked to the skin for hours and frozen with cold. Hours passed. Finally at 0300 we got the attack order.

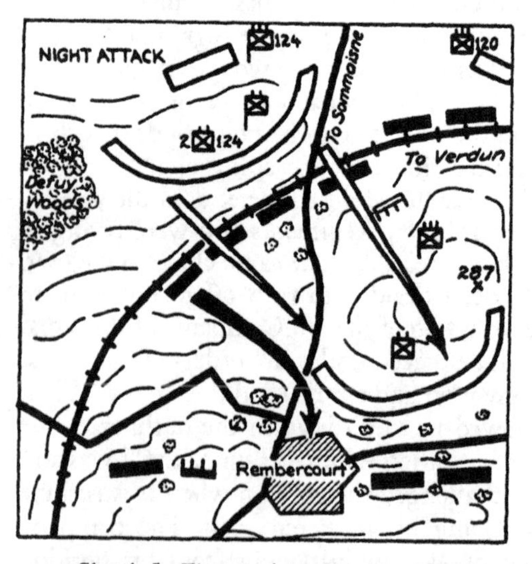

Sketch 5: The attack on Rembercourt.

In massed formation the battalion plunged down the slope onto the enemy along the railway, overran him, seized the cuts on the Sommaisne-Rembercourt highway, and stormed Hill 287. Wherever the enemy resisted he was dispatched with the bayonet, the rest of the battalion by-passing the local point of resistance.

With all four companies in line the battalion occupied Hill 287. No support had materialized on either flank, so both our flanks were refused. Units were badly mixed and reorganization proceeded slowly. Dawn began to break and the rain started to let up. The units were digging furiously to provide protection from the French artillery fire which was expected shortly. Work progressed very slowly in the wet, clayey soil. Over and over, shovels became coated with a thick, sticky coat of clay and had to be cleaned.

And now, in the gray morning light the shape of the hills around Rembercourt became plainly discernible; they dominated our new position. Suddenly our outpost sounded the alarm. Large masses of Frenchmen had been observed in the depression on the north side of Rembercourt.

At the moment of the alarm I was on the battalion right flank with Captain Count von Rambaldi's 6th Company. Closed columns of Frenchmen were marching into Rembercourt from the northwest. The 6th Company and parts of the 7th Company opened fire, and a very lively fire fight got under way at three to four hundred yards' range. Some French tried to find protection up the slope in the streets of Rembercourt, but the majority returned our fire. Most of our boys were so glad to have a Frenchman in their sights that they fired standing. After about a quarter of an hour the enemy fire slackened. In front of us at the north entrance to Rembercourt, there were large numbers of dead and wounded and our own zeal was responsible for the large gaps in our own ranks. The morning's fight was more expensive than the night attack.

We regretted not having been allowed to storm the village of Rembercourt and the hills on both sides of it. Our fighting spirit was unbroken in spite of all we had passed through and we wanted to be at grips with the French infantry which, so far, had proven inferior to us in all engagements.

After the fire fight died down all units continued to dig. Before they were a foot into the ground the French artillery let loose in its accustomed manner and prevented further work in the open.

So far the battalion staff had had little time to provide shelter for itself, the fight on Hill 287 and at the north entrance of Rembercourt having kept us constantly on the go. Now a French battery was firing at us from an uncovered position on the hill just west of Rembercourt. The range was scarcely more than eleven hundred yards. Fortunately there was a high percentage of duds because of the wet ground. We dived into plowed furrows to avoid the enemy's shells and covered ourselves with bundles of oats in hope of escaping the eyes of the enemy observers. The heavens opened again and our furrows turned into rivers. French shells hit near us, and our attempts to dig from the prone position were unsuccessful because of the clay coating on the shovel blades. We were literally covered from head to foot with a thick coat of sticky clay and were miserably cold in our wet clothes. In addition to all this my ailing stomach was particularly troublesome, and I was forced to change shell holes every half hour.

The attacks of our neighboring units had been stopped, leaving the 2d Battalion far ahead of the division front. At about 1000 a howitzer battery from the 45th Field Artillery attempted to help us from positions in the rear of our sector. The enemy maintained too great a superiority of fire, and the net results were to draw an even heavier fire on our own heads. Just as on the preceding days, we saw little of the French Infantry, which did not trouble us much in the way of fire.

Time stood still. A few months ago we would have laughed had we been told that this kind of miserable existence was even possible. We wanted to get out of our predicament and were not particular about methods. To attack would, of course, be preferable.

The French fire kept up all day long with countless shells thrown at our position on Hill 287. Just before dark we had the usual "goodnight kiss" and then in plain view, they hitched up and moved their batteries to the rear. They must have believed in maximum security at night.

Our losses on September 10 were considerable—four officers and forty men dead; four officers and 160 men wounded; eight missing.

After the night attack the French fortress of Verdun was all but encircled. A strip nine miles wide to the south of Verdun was all that separated the 10th Division east of Fort Troyon from the divisions of the XIII and XIV Corps attacking from the west. The only rail connection with Verdun was through the valley of the Meuse, and that was under German fire.

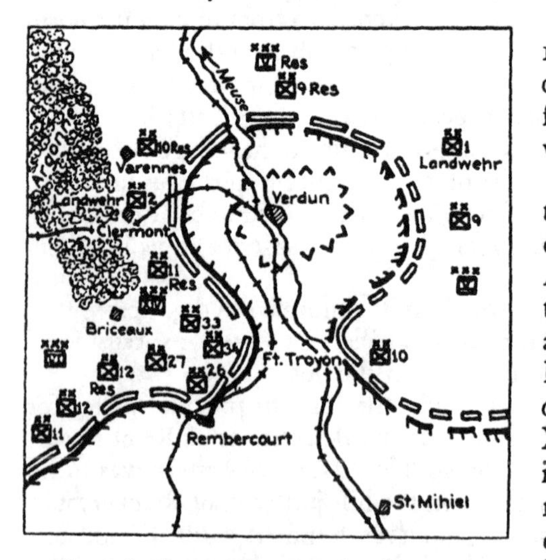

Sketch 6: The situation after the night attack of September 10-11, 1914

Night fell and we got busy with our digging. About midnight the kitchens arrived. The thoughtful Hänle had brought me dry clothes, underwear, and a blanket. Because of my stomach I decided to forego mess. I was not going to report sick as long as I could stand on my feet.

In dry clothes I got a few nightmare-packed hours and at daybreak returned to my pick and shovel.

On September 11, the French artillery carried on as before but our units were well entrenched and losses light. The continuous rains along with the cool temperature did not make our stay too pleasant. The kitchens again came up around midnight.

Observations: During a night attack it is very easy to fire on one's own people. In the 2d Battalion we missed doing this by a hair. The night attack of September 9 carried the 2d Battalion half a mile ahead of the division front, and we reached our assigned objective at a cost of few casualties. Had a continued advance been undertaken, it would have met with little resistance. The rain favored the attack. Heavy casualties only occurred when large masses of French were retreating into Rembercourt and while we were entrenching under French artillery fire. Had the French opened up before our trenches were a foot deep, losses would have been higher. A logical conclusion: Plenty of pick-and-shovel work before dawn. Because of ammunition shortage, our own artillery gave us but little support on September 10 and 11; and the French fired unmolested from exposed positions.

During action the volume of enemy fire was such that the kitchens came up only after dark. During the day they were several miles behind the front. The men quickly became accustomed to this manner of eating.

Chapter 3

FIGHTING NEAR MONTBLAINVILLE

I: Retirement Through The Argonne

At 0200 on September 12, I reported to the regiment for orders. A few hundred yards behind the 2d Battalion, in a poorly constructed trench covered with doors and boards, Colonel Haas issued his orders by candlelight: "Evacuate positions before daybreak; retire on Triancourt; 2d Battalion as rear guard holds the hills eleven hundred yards south of Sommaisne with the companies until 1100 and then follows the regiment."

On the one hand we were heartily glad to get out of this devil's cauldron, yet on the other we could not realize that we were to retreat. Certainly pressure on our front was not the reason. Verdun lay twenty miles to our rear and had but a single rail line connecting it to the rest of France. It seemed too bad to give it a breather. Well, the High Command had the big picture and must have had its reasons. Perhaps we were needed somewhere else.

Before dawn the 2d Battalion broke contact with the enemy. Our clothes were crusted over with a thick coat of dried mud and this, together with our lowered physical condition, made marching very hard. We left two companies as rear guard on the heights a mile and a quarter north of Rembercourt. At dawn, to our immense satisfaction and glee, the French artillery gave the empty positions a good pasting. This provided the Battalion wits with some material to work on.

We assembled in the wood west of Pretz and ran into our outposts at Triancourt. Captain Ullerich and I rode ahead to look over the situation. The rain was coming down in buckets again, and I was glad to be on horseback once more. The 5th and 7th Companies were assigned to the outpost, with the remainder of the battalion going into outpost reserve at Triancourt. After an afternoon inspection of the outpost, I returned to battalion headquarters and fell into a death-like sleep from which I could not be roused either by shaking or shouting. My battalion commander tried to wake me so that I could draw up my complete report. On September 13 I was called on the carpet because of this incident, but I had no recollection of their trying to awake me.

At 0600 on September 13 we were on our way back to the regiment. After passing Briceaux we headed through the Argonne. The sun shone brightly for the first time in days. Heavy supply columns had turned the roads into bottomless mud holes. The column halted at the entrance to the Argonne a mile north of Briceaux. Most of the artillery and trains were stuck in the mud, and double teams were required to get each piece and vehicle through. It was lucky for us that the enemy was not pressing the pursuit or shelling us with long-range artillery.

We were stalled for three hours. Moving over the soft forest road behind artillery that is continually getting stuck is more than exhausting. The troops were called on at frequent intervals to lend a hand on the wheels. It was evening before we reached Les Ilettes, where we had a brief halt for food and rest before continuing the march to the north through the Argonne. The twelve-hour march and the miserable road had nearly exhausted the troops, but in spite of this we continued the march through the dark night, our unknown objective apparently still far away. Exhausted men were beginning to drop out with increasing frequency. At every halt the men dropped in their tracks, and in a moment were sound asleep. Then when we were ready to move again, every one of them had to be shaken awake. We marched, halted, and marched again. I was constantly going to sleep and falling from my horse.

It was past midnight as we approached Varennes. The city hall was on fire—a fearful yet beautiful sight. I was given the mission of riding ahead to look for quarters in Montblainville, but the little town had only a few beds and no straw.

At 0630 on September 14 the silent and exhausted regiment stumbled through the dark streets. Billeting took little time, and within a few minutes Montblainville was again as silent as a grave. Everyone was fast asleep and quite unconscious of the hardness of his bed.

The same day, Major Salzmann took command of the battalion. In the afternoon we marched to Eglisfontaine, where we found cramped and dirty quarters. The battalion staff was quartered in a small verminous room but this was better than being out in the heavy rain which had just started again. My stomach was now in a terrible condition both day and night. I lost consciousness frequently.

During the succeeding days and nights, the French artillery bombarded all villages behind the front, including Eglisfontaine. We dug in near the town. On September 18 we marched to Sommérance to get a few days' rest. I was given quarters with a bed and I hoped to get my stomach in a little better shape. The opportunity to wash and shave, as well as to get a change of underwear, seemed the height of luxury.

During the first night we were alerted at 0400 and marched off to Fléville where the battalion went into corps reserve. There we stood in a pouring rain for three hours, after which we returned whence we came. On September 20 we had a real day of rest. Our men got their arms and equipment in shape.

Observations: Contact was broken off during the night of September 11 and 12 without the enemy's knowing it. Even on September 13 the enemy failed to pursue. Had he done so, our entry into the Argonne defile would have been most disagreeable. In the retreat on September 13 a march of twenty-seven miles was planned for troops who had been on outpost duty the previous night. The many halts and the assistance required by the bogged-down trains and artillery made this movement all the more difficult. The battalion was continually on the go for more than twenty-four hours.

II: Engagement At Montblainville; Storming Bouzon Woods

We were again alerted on the afternoon of September 21 and pulled out for Apremont where we were ordered to relieve a battalion of the 125th Infantry, then in the front lines on a ridge one mile west of Montblainville. The relief was to be accomplished after dark. The new position had little to recommend it. "A front slope position, all portions in view of the enemy, wet trenches, much rifle and artillery fire which caused casualties every day. Contact with the rear possible only at night."

In a pitch-black darkness, over loose soil, and again in a heavy downpour, we got across country guided by a detachment from the unit to be relieved. The relief was completed at midnight. The sector we had taken over consisted of short, discontinuous two-foot-deep trenches full of water. The garrison was lying a short distance to the rear with the men wrapped in their overcoats and shelter halves. We were told that the enemy was a few hundred yards in front.

The troops soon had the situation well in hand. Using mess kits they bailed the water out of their trenches and then started to deepen and improve them. Defuy Woods had taught them the value of trenches. Work went ahead rapidly in the soft ground, and at the end of a few hours most of the trenches were interconnected. The battalion could now look forward to the coming day without worry.

On September 22 the sun finally shone again. During the early hours all was peaceful in our sector. The enemy was some five to six hundred yards to our right on the edge of the Argonne forest. There was no sign of him along the Montblainville-Servon highway, which lay dead ahead of us. Off to the left the enemy occupied a bit of woods along the same highway. In spite of the relatively short range, we were able to move about outside the trenches without being fired on. Under these circumstances, the ripe plums on the trees close to our positions were quickly picked. Around 0900, a French battery began to lay its fire on our new trenches. Thanks to our spade work during the night, the casualties were light. Thirty minutes later the fire ceased, and during the next few hours we were subjected only to occasional harassing fire. Up to noon we saw nothing of French infantry, and a reconnaissance detachment left to find the location and strength of the enemy in the woods to our right.[1]

In the afternoon the kitchens arrived in the depression half a mile north of our position. In spite of the very lively harassing fire of all arms, carrying parties were able to get food clear up to the troops in the front lines.

At 1500 I went to the regimental command post in the vicinity of Point 180, about a mile northwest of Montblainville where I was given the situation and the attack order for the 2d Battalion. Strong enemy forces were located behind an abatis in the Boucon Woods along the Montblainville-Servon highway. All frontal attacks by the 51st Brigade on our right had failed. East of the Argonne and to our left the 1st Battalion, 122d Infantry, reinforced by the 1st Battalion, 124th Infantry, was attacking through Montblainville against the hills eleven hundred yards south of the town and was making good progress.

At twilight the 2d Battalion was to attack the enemy behind the abatis on the Montblainville-Servon road, take him in flank, and roll him up to the west. A neat but tough mission.

On my way back I studied the terrain carefully and considered the best means of carrying out our attack. To rush from our present position with the Montblainville-Servon road as initial objective seemed unattractive, for such an attack would not achieve surprise, would be subjected to flanking fire from the woods, and would cost us many casualties before reaching the highway. Finally, it would not place us on the Frenchmen's flank.

Having transmitted the regimental order, I made the following suggestion to the battalion commander: First, that we evacuate the positions on the hill a mile west of Montblainville and assemble the battalion on the north slope of the hill where there was some cover; then, in a formation of considerable depth, advance up the draw to the east of our present position and seize the small woods seven hundred yards west of Montblainville.

This small woods had been shelled by our artillery a short time before and to all appearances was deserted. It was possible on account of the topography to execute our movements without being observed by the enemy.

Once in the woods, the Battalion could deploy to the west preparatory to an attack south of the highway against the eastern edge of the

[1]Edited out of the text at this point was a charge made by Rommel that the French infantry deliberately killed wounded Germans lying on the battlefield. ED.

Argonne. Such an attack would strike the flank of the enemy lying along the Montblainville-Servon highway. If we moved out immediately we could still attack in the twilight.

My proposal was translated into action. By going back a man at a time, platoon after platoon managed to evacuate the positions on the south slope. A few men were slightly wounded by lively rifle fire. The whole battalion was soon assembled on the north slope. The enemy continued to fire at the empty positions. We advanced in column of files, the battalion staff leading toward the small woods seven hundred yards west of Montblainville. The Frenchmen, oblivious of our departure, were still pasting our empty positions.

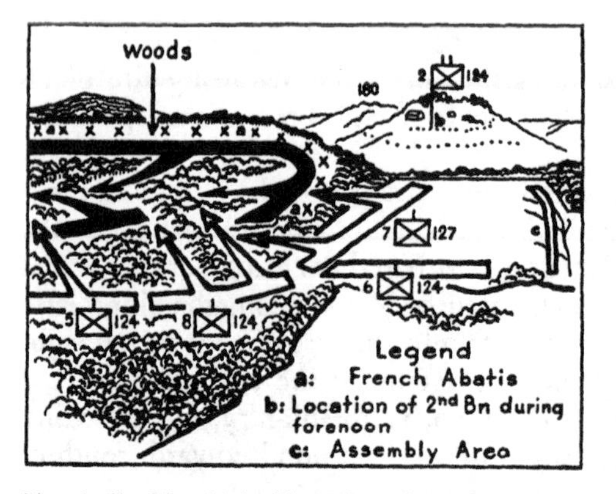

Sketch 7: Montblainville. View from the south.

We reached the little woods without being detected. A shallow rifle trench ran along the north edge where we found a few pieces of equipment such as packs, canteens, and rifles. The former occupants had probably deserted the trench during the afternoon because of German artillery fire. We deployed to the west and prepared to attack the enemy at the edge of the woods. He seemed unaware of our presence; at least we had not been fired on from that direction.

Our objective was up a slope about a quarter of a mile away. Six hundred yards south of the highway, we located a good, covered avenue of approach leading directly into the position, and the 5th Company was brought up this draw to within a hundred yards of the edge of the woods. In the meantime the 7th and 8th Companies assembled between the highway and the draw. The 6th Company constituted the battalion reserve. The staff was up with the 5th Company. The companies were given their orders. Our plan was to surround the enemy established along the highway. The formation was echeloned to the left.

It was getting quite dark when Major Salzmann gave the attack signal. The approach was made without noise, and soon the leading files of the 5th Company reached the forest. The 7th and 8th Companies were about three hundred yards from the edge of the woods. There was no sign of the enemy. His attention appeared to be focused on our old positions north of the highway.

The 5th Company continued on through the underbrush and together with the battalion staff was soon lost among the trees. Suddenly the 7th Company ran into the enemy along the highway, and a short fire fight got under way at ranges of a hundred yards or less. The 5th Company and the battalion staff turned right, the 8th Company and the left of the 7th Company turned half-right and then the whole battalion charged ahead.

The abatis along the road lost its value and our unexpected attack in flank and rear caused panic to spread among the French defenders, who fled to the west. Darkness put an end to the fighting. Our booty included fifty prisoners, several machine guns, ten artillery ammunition wagons as well as the hot meal that was being cooked at the time of our attack. Our losses were: Lieutenant Parst and three men killed. An officer and ten men wounded. There were also these additional results from our attack: the panic spread to an entire French brigade and it abandoned a strong position without firing a shot. During the night the 51st Württemberg Brigade gathered in a large number of fugitives most of whom were taken at the crossroads of the Montblainville-Servon road and the old Roman road. (See sketch 8.)

We bivouacked on the field and were very cold. Luckily captured French supplies gave our horses enough to eat.

On September 23 at daybreak, I accompanied Colonel Haas on a reconnaissance as far as the old Roman road. After this the 2d Battalion received orders to move south along the eastern edge of the Argonne Forest as far as Les Escomportes farm. While I was still at the regimental CP, my battalion, contrary to orders, marched off through the woods and I could find no trace of them. I tried to get down to Les Escomportes farm by going along the eastern edge of the woods, but noticed that the French still occupied the place and that they had machine guns. I did not find the battalion until afternoon, by which time it had advanced through the woods, by-passed Les Escomportes farm, reached the hill eleven hundred yards south of the farm, and had driven off the few outposts found there. By the time I caught up, the French were using artillery on us again. It is a mystery to me how they obtained our location in the middle of a woods and how they managed to deliver effective fire in so short a time.

The hungry and tired men lay down under trees and in some improvised shelter the French had built out of branches. They had gone without food since early morning, and I rode back to get the kitchens which were located in the vicinity of Apremont. I located them half a mile north of Montblainville. It turned out, however, that the horses were in no condition to negotiate the soft ground between there and the outfit. They got stuck a quarter of a mile east of Les Escomportes farm, and the units were fed piecemeal between midnight and 0300.

In the meantime a regimental order had been received ordering us to be at Les Escomportes farm by 0500; consequently, we got little sleep.

Observations: The relief of a front-line battalion at night—Guides must lead the way. The relief must be noiseless, otherwise the enemy can break up the movement and cause needless casualties simply by opening fire.

Again the 2d Battalion made good use of the spade before daybreak and survived the enemy artillery bombardment with but few casualties.

Combat reconnaissance.—It is advisable to have strong fire support ready for such a reconnaissance as was made during the forenoon of September 22. Losses are thus avoided. Under certain circumstances it is advisable to furnish a light machine gun for fire support.

On September 22 the 2d Battalion succeeded in evacuating a position on a forward slope in daylight with few casualties in spite of the fact that the enemy was only six hundred yards away. Men withdrew singly. In my opinion, such a maneuver would be possible today. Of course, the enemy would have to be held down by artillery and heavy infantry weapons. Also, the employment of smoke would make such a movement easier.

The evening attack of the 2d Battalion in the flank and rear of the enemy, who was strongly entrenched in the Argonne Forest, was a great success and cost us little in casualties. We were able, thanks to the terrain, to attack with our left company well out, and the formation paid off when we made contact, for we were then able to turn the right flank. The spread of panic among the whole brigade resulted in our capture of the entire position.

The night of September 23-24 provides a good example of ration supply difficulties in a war of movement.

III: Forest Fighting Along the Roman Road

As ordered, the 2d Battalion arrived at Les Escomportes farm at 0500 on September 24. We halted and rested. In a small, dark farmhouse room, Colonel Haas ordered the Salzmann Battalion to move through the forest and seize and hold the intersection of Four-de-Paris—Varennes road and old Roman road.

The stimulation brought on by anticipation of a new mission caused us to overlook our fatigue, and I even forgot my complaining stomach.

While the battalion was getting under way the sun rose out of the morning mists like a red ball of fire. Maintaining direction by compass, we plowed through the dense and trackless underbrush toward the intersection. I marched on foot at the head of the column which was frequently forced to detour around impenetrable undergrowth. During the last few years of peace the junior officers of the 124th Infantry had been given intensive training in the use of the compass at night, and this training now reaped its just reward.

It took us an hour to reach the old Roman road at a point about two-thirds of a mile from our objective. We moved off to the south with the usual march security and the staff marched behind the advance guard.

A mounted reconnaissance detachment returned from the Four-de-Paris—Varennes highway and reported that the enemy had entrenched himself along the road. Caution was indicated. Preceded by security elements, the 5th and 6th Companies headed toward the highway over different routes. The tall trees were now clearly visible, but the underbrush was just as thick as before. The battalion commander remained with the 7th and 8th Companies in the vicinity of the shelter hut while I went on with the point of the 6th Company. A few dead Frenchmen lay beside the road. Suddenly we heard the sound of rapidly approaching horses up ahead. Were they friend or foe? Along this badly overgrown road the maximum range was eighty yards at the most. The point dove into the bushes on both sides of the road. The next moment a herd of riderless horses came galloping around the turn, stopped at sight of us, and dashed off to the right.

While the 6th Company reached the main road without further incident, the 5th Company, over on the left became engaged in a lively scrap.

I galloped back to the battalion to report. At the same time the 5th Company reported that they had run into the enemy behind the abatis six hundred yards south of the shelter hut, that he was being reinforced and that a further advance was impossible without additional support. Soon after this, two badly-wounded 5th Company officers were brought back. The volume of firing increased on the 5th Company front and shots were also heard from the 6th Company. Bullets crashed through the forest and we could not tell whether or not they were from snipers.

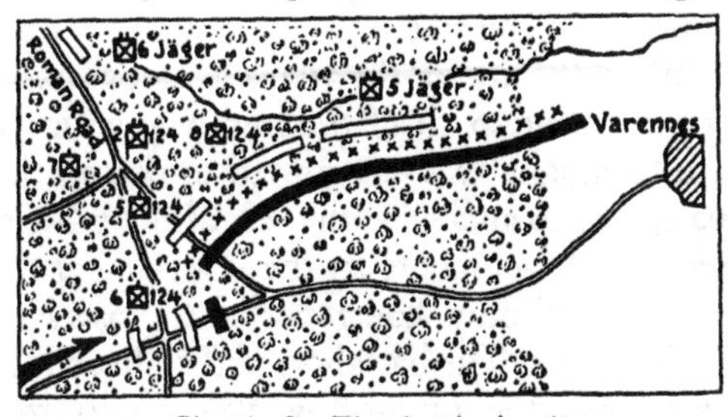

Sketch 8: The battle in the woods, September 24, 1914.

Major Salzmann committed the 8th Company

on the left of the 5th Company. Both companies were to attack simultaneously and throw the enemy back across the Four-de-Paris—Varennes highway.

The 8th Company had barely left when leading units of the 5th and 6th Jäger Battalions arrived at the shelter hut. We learned that their mission was identical to ours. After a little deliberation, Major Salzmann committed the 5th Jäger Battalion on the left of the 5th and 8th Companies, giving the Jägers the mission of assisting our companies in throwing the French back across the road.

Within forty-five minutes this attack had been stopped cold. According to the reports of many wounded, the enemy was located in considerable strength behind an abatis and had many machine guns in position.

About this time Captain Count von Rambaldi, of the 6th Company, came back slightly wounded and reported that he was opposed by a company of French on the Four-de-Paris—Varennes highway about two hundred yards east of his present position and that the woods to the west of his unit still had not been cleared. I went up to the 6th Company to look over the situation. Advancing south of the Four-de-Paris—Varennes highway with a strong reconnaissance detatchment from the 6th Company, I ran into the enemy sixty yards east of the 6th Company's hedgehog position. As a result of this reconnaissance I believed that we were opposed merely by a strong enemy outpost.

On returning to the battalion, I recommended that we attack along both sides of the road to seize Varennes, the 6th Company going straight down the road while the 7th Company and the 6th Jäger Battalion attacked on either side of the road. This maneuver would take the enemy now holding up the others in flank.

Before any action transpired, we received a regimental order to clear the Varennes highway. The 5th and 6th Jäger Battalions were attached to the 2d Battalion for this operation. At this same time the 6th Company reported that closed columns of French were approaching from the direction of Four-de-Paris; so it was high time to clear up the situation in the east.

We got ready for the attack as fast as possible. The 6th Jäger Battalion was to move south of the highway with its left flank on the road; the 7th Company was committed to the north of the road. The 6th Company was to attack on the left of the 7th Company after leaving strong security detachments on the Four-de-Paris road.

When all units had reported themselves ready, they moved forward to the attack. The battalion staff followed the 7th Company. A hundred yards from the jump-off we were forced to the ground by heavy enemy fire. We could hardly see more than twenty-five yards through the thick undergrowth and could see nothing of the enemy. Our companies opened fire and worked toward the invisible enemy by means of short rushes. Because of the deafening sound of the rifles, it was impossible to approximate the distance to the enemy. His fire increased in intensity. Our attack was halted.

In order to get the 7th Company moving forward again, Major Salzmann and I got into the front line. I took a rifle and ammunition from a wounded man and took command of a couple of squads. It was impossible to handle a larger unit in those woods. Several times we rushed through the bushes toward the enemy whom we supposed to be very close. We never succeeded in getting to him, but time and again his rapid fire forced us to the ground. The calls for aid men told us that our casualties were increasing.

Pressed flat on the ground, or behind thick oak trees, we let the enemy fire go by and then at the first let-up, attempted to gain more ground in his direction. It was becoming harder to get the men to move forward; consequently we gained ground slowly. Judging from the sound of the fighting, our neighbors were about abreast of us.

Once again we rushed the enemy in the bushes ahead of us. A little group of my former recruits came with me through the underbrush. Again the enemy fired madly. Finally, scarcely twenty paces ahead I saw five Frenchmen firing from the standing position. Instantly my gun was at my shoulder. Two Frenchmen, standing one behind the other, dropped to the ground as my rifle cracked. I still was faced by three of them. Apparently my men sought shelter behind me and couldn't help me. I fired again. The rifle missed fire. I quickly opened the magazine and found it empty. The nearness of the enemy left no time for reloading, nor was any shelter close at hand. There was no use thinking of escape. The bayonet was my only hope. I had been an enthusiastic bayonet fighter in time of peace and had acquired considerable proficiency. Even with odds of three to one against me, I had complete confidence in the weapon and in my ability. As I rushed forward, the enemy fired. Struck, I went head over heels and wound up a few paces in front of the enemy. A bullet, entering sideways, had shattered my upper left leg; and blood spurted from a wound as large as my fist. At any moment I expected a bullet or bayonet thrust. I tried to close the wound with my right hand and, at the same time, to roll behind an oak. For many minutes I lay there between the two fronts. Finally my men broke through the bushes and the enemy retreated.

Lance Corporal Rauch and Private Rutschmann took care of me. A coat belt served as tourniquet and they bandaged my wound. Then they carried me back to the hut in a shelter half.

From up ahead word came that the enemy, leaving two hundred prisoners, had been driven from behind his abatis and out of the woods. Our own casualties had been considerable: 30 dead, including 2 officers, and 8 wounded, including 4 officers for the 2d Battalion alone. As it was later reported in the regimental history, this was the third time in three days that the battalion had distinguished itself.

It was tough leaving these brave men. As the sun set, two men carried me back to Montblainville in a shelter half attached to two poles. I felt but little pain, yet I fainted from loss of blood.

When I regained consciousness in a Montblainville barn, Schnitzer, the battalion surgeon, was working over me. Hänle had brought him. My wound was dressed again, and I was loaded into an ambulance beside three wounded, groaning comrades. We left for the field hospital, the horses trotting over the shell-torn road; and the jolting which resulted caused me great pain. When we arrived around midnight one of the men beside me was already dead.

The field hospital was overcrowded. Blanketed men lay in rows along the highway. Two doctors worked feverishly. They re-examined me and gave me a place on some straw in a room.

At daylight an ambulance took me to the base hospital at Stenay where, a few days later, I was decorated with the Iron Cross, Second Class. In the middle of October after having had an operation, I was taken home in a private car that had been placed at the Army's disposal.

Observations: The enemy along the Four-de-Paris—Varennes highway made it most difficult for the 2d Battalion to accomplish its mission. Three battalions were finally engaged in the attack in the woods, and it was only after sustaining heavy losses that they were able to drive the enemy from the dense woods.

The high casualty rate began with the opening of the engagement. Among others, three officers were lost. It is hard to say whether or not French tree snipers were at work, for none was discovered and brought down.

With our high casualty rate we had difficulty in getting the men to drive ahead. In forest fighting, the personal example of the commander is effective only on those troops in his immediate vicinity.

In a man to man fight, the winner is he who has one more round in his magazine.

Part Two

TRENCH WARFARE
IN THE ARGONNE AND HIGH VOSGES

★ ★ ★ ★ ★ ★ ★ ★ ★ ★ ★ ★ ★ ★ ★

Chapter 4

ATTACK IN THE CHARLOTTE VALLEY

Shortly before Christmas I was released from the hospital, but my wound had not healed and hampered my walking. Service in a replacement battalion was distasteful so I returned to my outfit.

In the middle of January, 1915, I joined the regiment in the western part of the Argonne. The bottomless road from Binarville to the regimental CP was indicative of conditions in the Argonne Forest. I assumed command of the 9th Company which needed a commander. A narrow, corduroy footpath led forward from the regimental CP for a distance of about half a mile. Occasional rifle bullets flew through the winter woods and a few shells whistled overhead, forcing me to dive for cover in the deep clayey communications trench. By the time I arrived at my company CP my uniform had lost the telltale marks of the soldier returning from leave.

I assumed command of about two hundred bearded warriors and a 440-yard company sector of the front line. A French reception committee greeted me with a concentration of "Whiz Bangs." The position consisted of a continuous trench reinforced by numerous breastworks. Several communication trenches led to the rear. Shortages in barbed wire prevented the erection of obstacles out in front. In general the position was poorly developed, and surface water had kept the trench depth to three feet or less in some places. The dugouts, built to accommodate from eight to ten men, were of necessity equally shallow, and their roofs stuck out above the ground level making them excellent targets. Their roofs were nothing more than a couple of layers of thin logs which at best were only splinterproof. During the first hour of my command a shell landed smack on one of them and severely wounded nine men. My first order was that whenever artillery opened on us all dugouts would be vacated and the men would take cover in the trench proper. I also issued orders that the dugout roofs would be strengthened so that they could at least withstand field artillery fire. This work started at dark. Several large oak trees near our position proved to be dangerous to our safety. Whenever shells burst against them they deflected the fragments straight into our trenches; so I ordered several of them chopped down.

Stimulated by my new command, it was not long before I was my old self again. For a 23-year old officer there was no finer job than that of company commander. Winning the men's confidence requires much of a commander. He must exercise care and caution, look after his men, live under the same hardships, and—above all—apply self discipline. But once he has their confidence, his men will follow him through hell and high water.

Each day brought plenty of work. We lacked boards, nails, clamps, roofing paper, wire, and tools. The headquarters dugout which I shared with a platoon commander was four and a half feet high and contained a table and cot made of beach sticks tied together with wire and string. The walls were bare earth, and water trickled down constantly. During wet weather, water also leaked through the roof, which was made of two layers of oak trunks and a thin layer of earth. Every four hours the dugout had to be bailed out to prevent flooding us out. We built fires only at night, and in damp winter weather were cold all the time.

We could not see anything of the enemy position across from us because of the thick underbrush. The French were in better shape than we. They did not have to cut trees for lumber, for they received all the necessary materials from their supply dumps. Their location in extremely thick woods and our shortages in artillery ammunition limited the amount of harassing fire to which they were subjected. The enemy positions were some three hundred yards away on the other side of the small valley. To hinder our work parties, the enemy frequently sprayed us with small arms fire. Unpleasant as this was, we disliked the "Whiz Bangs" even more because of the short time interval between their discharge and impact. Whoever was caught in the open by one of these shells hit the dirt immediately if he hoped to avoid being hit by the shell fragments.

Toward the end of January 1915 it rained and snowed on alternate days; and from January 23 to 26 the company went into reserve about five hundred feet behind the front line. There the dugouts were still worse, hostile artillery fire more troublesome, and daily losses equal to those up front. The company was used for service work: *i.e.,* transport of materials, construction of dugouts, improving communication trenches, and laying out corduroy footpaths. We were glad when the time came for us to go forward again. Morale was high and officers and men alike were willing to endure any hardship in order to defend our native land and achieve the final victory.

On January 27 two of my men and I went on reconnaissance which took us up a trench leading toward the enemy from the left of my company sector. At this time we were located in an old French position which had been taken on December 31, 1914. After removing some obstacles in the trench we proceeded cautiously; and about forty yards down the trench we came upon some dead Frenchmen who had probably been lying unburied between the fronts since the attack. To the left of the trench was a small graveyard and, at the end some hundred yards from our own position, a deserted medical aid station which, located in the deepest depression between the lines, was well dug in, well sheltered, and capable of holding twenty men. During the tour we saw nothing of the enemy although he delivered his usual harassing fire against our positions. Judging from the sound of his weapons he was about five hundred feet away on the other side of the valley. I decided to turn the dugout into an advanced strongpoint, and we started work that same afternoon. From this position we could even hear the French talking across the way. I did not believe it wise to send any scouts forward, for they would have had too hard a time getting through the dense underbrush without being seen, and they would have been shot before obtaining any worthwhile information.

In order to pin a maximum of the enemy strength in the Argonne, small diversionary attacks were ordered for January 29, 1915; and all regiments of the 27th Division were to participate. Following the blowing of a French mine shaft, our regiment was to conduct a heavy raid in the 2d Battalion sector. While the raid was progressing, artillery would open up and pin the enemy in front of the 3d Battalion. For this purpose a howitzer battery from the 49th Field Artillery was made available and given time to complete its registration fire. During the operation the 10th Company would have to shift, while the 9th Company was not to advance but to cut off all enemy attempts to escape on the flanks.

January 29 dawned cold with the ground frozen. At the start of the operation I was up forward in our new strongpoint with three rifle squads. We were a hundred yards ahead of our positions and heard our own shells whistling overhead, some striking the trees, others landing to our rear. Then they blew the mine; and earth, sticks, and stones rained on the landscape. Hand-grenade blasts from the right and intense small-arms fire followed the explosion. A lone Frenchman ran up to our position and was shot.

A few minutes later the adjutant of the 3d Battalion came up, reported that the attack on the right was going well, and said the battalion commander wished to know if the 9th Company cared to join in the fun. To accept was a distinct pleasure.

I realized that I could not move my company from our trenches in deployed formation, for enemy artillery and machine guns had our range and any advance on our part would be reported by his treetop observers. To avoid this I had my men crawl up a trench that extended to the front from the right of our position. After they reached the end of the trail, they deployed to the left; and after about fifteen minutes the company was assembled in an area a hundred yards in front of our position and on the slope leading down to the enemy. Carefully we crawled through the bare underbrush toward the enemy; but before we could reach the hollow he opened on us with rifle and machine-gun fire that stopped us cold. There was no cover, and we could hear the bullets slam into the frozen ground. Up ahead a few oak trees sheltered a handful of my men. I could not locate the enemy even with my binoculars. I knew that to remain where we were would cost us dearly in casualties, for even though the enemy fire was unaimed, it made up in volume what it lacked in direction. I wracked my brains to find a way out of this mess without suffering too great losses. It is in moments like this that the responsibility for the weal or woe of one's men weighs heavily on a commander's conscience.

I had just decided to rush for the hollow sixty yards ahead, since it offered a little more cover than our present location, when we heard the attack signal far off to the right. My bugler was right beside me and I had him sound the charge.

In spite of the undiminished volume of fire directed at us, the 9th Company jumped up and, cheering lustily, dashed forward. We crossed the hollow and reached the French wire entanglements only to see the enemy hurriedly abandon his strong position. Red trousers flashed through the underbrush and blue coat tails were flying. Totally oblivious of the booty left behind in the abandoned positions, we rushed after them. By sticking close to the enemy's heels we managed to smash through two other defensive lines which had been well provided with wire entanglements. At each position the enemy ran before we got to him. As proof of the meager resistance, we had no losses whatever. (See sketch 9.)

We passed over a height and the woods began to thin out. We could see the enemy running before us in a dense mass, so we pounded after him, shooting as we went. Some of the company cleaned out the dugouts, and the rest of us kept going until we reached the edge of the woods six hundred yards west of Fontaine-aux-Charmes. At this point we were half a mile south of our initial position. Here the terrain sloped down again, and the fleeing enemy had disappeared in low undergrowth. We had lost contact on both flanks and to the rear and on both sides we heard the sounds of a bitter struggle. I assembled the company and occupied the edge of the woods west of Fontaine-aux-Charmes and then tried to reestablish contact with the adjacent units. To the accompaniment of general laughter, a soldier brought some articles of feminine wearing apparel out of a dugout.

A reserve company arrived, and after giving it the job of reestablishing contact, we moved off down hill to the southwest through the light shrubbery of this sector where the terrain had been largely cleared of troops. My unit advanced in a column behind strong security elements. We had just crossed a hollow when strong fire from our left forced us to the ground, but the enemy could not be seen. In order to maintain our impetus, we moved off to the west, bypassed the hostile fire, and then resumed our advance to the south through open woods.

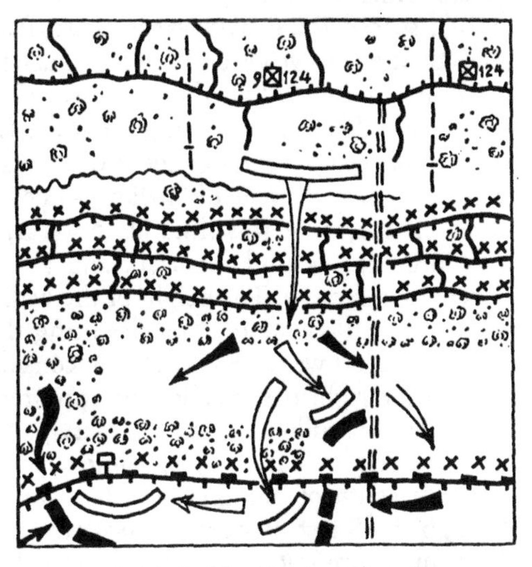

Sketch 9: The attack against the "Central" position, January 29, 1915.

At the upper edge of these woods, we ran into a wire entanglement the like of which we had never seen. It was more than a hundred yards wide and stretched out to the flanks as far as the eye could see. The French must have cut down the whole forest here. I could see three of my men waving at us from the far side of the wire, and I concluded that the enemy had yet to occupy the strong position. That being the case, the smart thing to do was to hang on until reserves came up.

I tried to move on down the narrow path that led through the wire, but enemy fire from the left forced me to hit the dirt. The enemy was nearly a quarter of a mile away and certainly could not see me because of the density of the wire, yet ricochets rang all around me as I crawled through the position on all fours. I ordered the company to follow me in single file, but the commander of my leading platoon lost his nerve and did nothing, and the rest of the company imitated him and lay down behind the wire. Shouting and waving at them proved useless.

This position, constructed like a fortification, could not be held by three men alone, and the company had to follow. By exploring the west, I found another passage through the obstacle and crept back to the company where I informed my first platoon leader that he could either obey my orders or be shot on the spot. He elected the former, and in spite of intense small-arms fire from the left we all crept through the obstacle and reached the hostile position.

To secure the position, I had my company deploy in a semi-circle and dig in. The position was called "Central" and was constructed in accordance with most recent design. It was part of the general defensive system which ran through the Argonne and consisted of strong blockhouses, spaced some sixty yards apart, from which the French could cover their extensive wire entanglements with flanking as well as frontal machine-gun fire. A line of breastworks connected the individual blockhouses, and this wall was so high that fire from the fire step could reach any part of the wire entanglement within range. The wall was separated from the entanglement by a ditch some fifteen feet wide which was water-filled and, at this time of the year, frozen over. Deep dugouts were provided behind the wall, and a narrow road ran along some eleven yards from it. The height of the wall was such as to offer concealment and defilade to any vehicles using the road.

From the left we were subjected to considerable small-arms fire, while over on the right the installations appeared to be unoccupied. Around 0900 I sent the following written message to my battalion:

"9th Company has occupied some strong French earthworks located one mile south of our line of departure. We hold a section running through the forest. Request immediate support and a resupply in machine-gun ammunition and hand grenades."

Meanwhile the troops were trying to make an impression on the frozen ground with their spades, but it was only by using the few available platoons and crowbars that we made any progress. We had been working for some thirty minutes when the left outpost reported that the enemy was retiring through the wire some six hundred yards to the east in closed column. I had one platoon open fire. Part of the enemy headed for cover, but others who were still north of the obstacle turned farther to the east and apparently reached the covered road behind the works, for very shortly after our opening fire we were attacked from that direction.,

Our digging had produced meager results, and, in studying the situation, I noticed a bend in the position in the vicinity of Labordaire. This bend would make an excellent strongpoint to hold if we were to retain our foothold in the enemy position. My company fought its way to this new position where we quickly found shelter behind tree trunks and began returning the enemy fire in sufficient volume to keep him some three hundred yards away, and he soon began to dig in. Shortly after this the fire slackened and soon died.

My foothold included four blockhouses, my company being deployed in a semicircle with a platoon of fifty men in concealed reserve between the wire entanglement and the position. Here another narrow zigzag passage led through the wire field. Time passed and we began getting anxious about our reinforcements and supplies. Suddenly reports from the right indicated that more French were retreating through the wire some fifty yards from us. The platoon leader wanted to know if he should open fire. What else was there for us to do? We were about to get into a nasty scrap, and there was no use allowing the French to start it free from casualties. If we fired at once, then the French would turn to the west and get into position through the next passage; it was also possible that they might get across our line of communication and so surround us. I opened fire.

From the high French breastworks rapid fire struck the nearby enemy, and a bitter struggle developed, with the French fighting bravely. Fortunately most of the new enemy, estimated as one battalion, turned off to the west, traversed the wire entanglements 350 yards away, and moved toward us from the west on a broad front. The ring about the 9th Company closed leaving but one narrow path through the wire to connect us with the battalion. Even that lifeline was swept by enemy fire from the east and west. On the right our heavy fire kept the enemy pinned to the ground, but the enemy on the left had made progress and was getting dangerously close. Ammunition was getting scarce, and I stripped the reserve platoon of most of its equipment. I decreased the rate of fire in order to conserve ammunition as long as possible, but the enemy on the west kept crawling closer. What was I to do once my

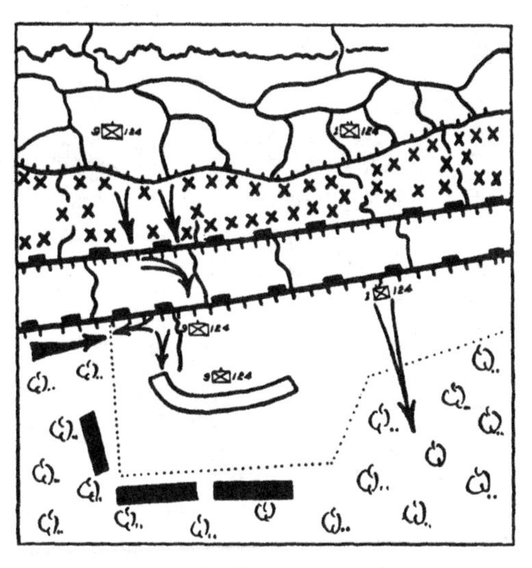

Sketch 10: The attack against the "Central" position.

ammunition was exhausted? I still hoped for help from the battalion. Minutes seemed like hours.

A fierce battle raged around the blockhouse on the extreme right, and we expended our last grenades in its defense. A few minutes later, about 1030, a French assault squad succeeded in taking it and used its embrasures to pour rifle and machine-gun fire into our backs. This report reached me at the same moment that a battalion order was shouted across the entanglement by a runner: "Battalion is in position half a mile to the north and is digging in. Rommel's company to withdraw, support not possible." Again the front line was calling for ammunition, and we had enough for only ten more minutes.

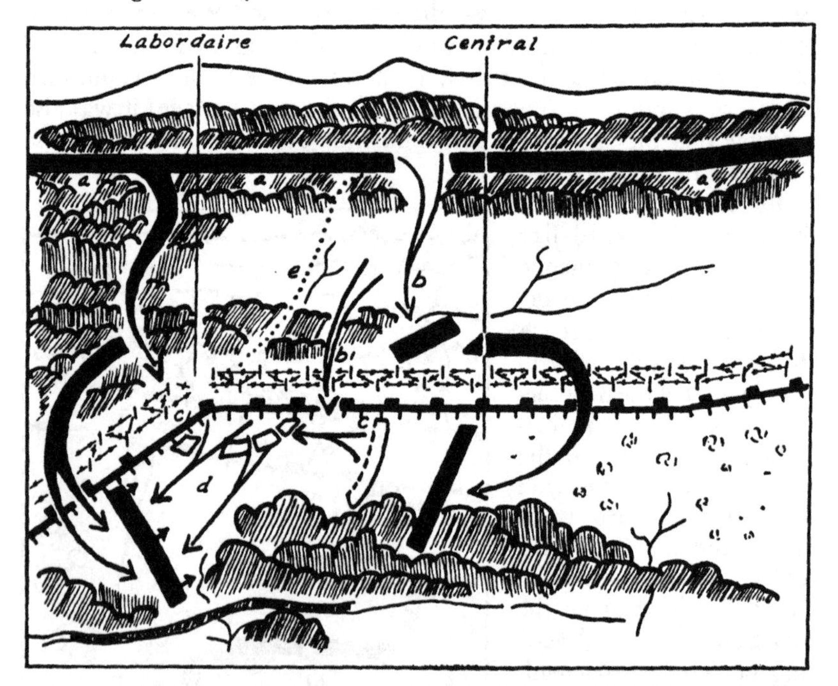

Sketch 11: The attack against the "Central" position, January 29, 1915. View from the south. (a) Third French position. (b) 9th Company exploits the breakthrough and penetrates as far as the "Central" position. (c) 9th Company holds portions of "Central" and Labordaire positions. (d) Attack prior to breaking off combat. (e) Route followed during retirement.

Now for a decision! Should we break off the engagement and run back through the narrow passage in the wire entanglement under a heavy cross fire? Such a maneuver would, at a minimum, cost fifty per cent in casualties. The alternative was to fire the rest of our ammunition and then surrender. The last resort was out. I had one other line of action: namely, to attack the enemy, disorganize him, and then withdraw. Therein lay our only possible salvation. To be sure, the enemy was far superior in numbers, but French infantry had yet to withstand an attack by my riflemen. If the enemy in the west were thrown back, we would have a chance of getting through the obstacle and only have to worry about the fire of the more distant enemy on the east. Speed was the keynote of success, for we had to be gone before those we had attacked could recover from their surprise.

I lost no time in issuing my attack order. Everyone knew how desperate the situation was, and all were resolved to do their utmost. The reserve platoon drove to the right, recapturing the lost blockhouse and carrying the whole line along with its impetus. The enemy broke and ran. With the French running away to the west, the proper moment to break off combat had come. We hurried eastwards and negotiated the wire entanglement in single file as fast as possible. The French on the east opened up on us, but a running target was not too profitable at a range of three hundred yards. Even so, they got a few hits. By the time the enemy on the west had recovered and returned to the attack, I had the bulk of my outfit on the safe side of the wire. Aside from five severely wounded men who could be taken along, the company reached the battalion position without further incident.

The battalion, with my company on the left, was established in the dense forest directly south of the three occupied French positions. The 1st Battalion was having trouble and was out of direct contact with our left, but by means of liaison squads we managed to keep in touch with their right. My company dug in some hundred yards from the forest edge. Digging in the frozen ground was no fun.

So far the French artillery had devoted its entire attention to our old position and to the rear areas; and during the attack we had been spared its attention, probably due to poor infantry-artillery liaison. This had been remedied now, and we were subjected to a very heavy volume of retaliatory fire which interfered with our digging since the forward edge of the forest received particular attention. I prepared my report of the morning's activities on a message blank and accompanied it with a sketch.

Late in the afternoon, following a heavy artillery preparation, the enemy counterattacked. Masses of fresh troops stormed through the underbrush only to be met by our small-arms fire. They fell, sought cover, and returned our fire. Here and there a small group tried to work its way closer, but in vain! Our defensive fire smothered the attack with heavy losses, and large numbers of dead and wounded lay close to our lines. Under cover of darkness the French withdrew to the edge of the forest one hundred yards away and dug in.

The infantry fire died down, and we too began to dig, for our own trenches were only twenty inches deep. French artillery again interrupted this work and sharp-edged shell fragments whistled about our ears, struck, and destroyed trees as if they were matches.

Our positions offered inadequate cover for the harassing fire which, with few breaks, kept up all night. Wrapped in overcoats, shelter halves and blankets, we lay shivering in the shallow trench. I could hear the men jump as each new concentration hit near us. During the night we lost twelve men, which was a heavier loss than we had sustained during the entire attack. No rations could be brought up.

At dawn the hostile artillery activity slackened and we began to work on deepening our positions; but we were not allowed much time. At 0800 artillery fire forced us to quit, and the fire was followed by a strong infantry attack which we threw back with little difficulty. The same fate met succeeding attacks, and by afternoon our positions were deep enough so that we could stop worrying about the effects of artillery fire. We had no communication trenches to the rear; so we had to wait until dark for our first hot meal.

Observations: The attack on January 29, 1915, showed the superiority of the German infantry. The attack of the 9th Company was no surprise, and it is difficult to understand why the French infantry lost its nerve and abandoned a well-prepared defensive position lavishly protected by wire, three lines deep, and well-studded with machine guns. The enemy knew the attack was coming and had tried to stop it by means of heavy interdiction fire. The fact that we were able to resort to offensive action and break from the encircled Labordaire position is ample proof of the combat capabilities of our troops.

It was unfortunate that neither the battalion nor the regiment was able to exploit the 9th Company's success. With three battalions in line, inadequate reserves were available. Shortages in small-arms ammunition and hand grenades increased our troubles in the defense of Labordaire. Several things happened simultaneously to render our situation most critical: First, the enemy seized the blockhouse on the extreme right; second, we received the battalion order to withdraw; third, we were short of ammunition; and, finally, our way back through the wire was swept by enemy fire. Any decision, other than the one made, would have resulted in terrific casualties if not total annihilation. Above all, it was impossible to wait for darkness; for the last round would have been fired well before 1100. Attacking the weaker enemy force on the east would not have paid dividends, for the more aggressive attack came from the west; and attacking to the east would have given the western force an excellent opportunity to strike us in rear. Breaking off the fighting in Labordaire confirms the statement in the *Field Service Regulations:* "Breaking off combat is most easily accomplished after successful offensive maneuver."

In making our hasty preparations for the attack, we gave no thought to heavy entrenching tools. The solidly frozen ground made our light tools almost useless. Even in the attack the spade is as important as the rifle.

Although there was a better field of fire from the edge of the forest, the new position was one hundred yards inside the woods. We had no intention of exposing the troops to a repeat performance of the Defuy woods bombardment, and still had a field of fire good enough to repel several French infantry attacks with heavy losses.

The losses from hostile artillery fire during the night of January 29-30 were so heavy because the troops did not dig in to a proper depth.

Chapter 5

TRENCH FIGHTING AT "CENTRAL" AND IN THE CHARLOTTE VALLEY

I: Trench Warfare in the Argonne

Our new positions were an improvement. We were located at a higher elevation, and surface water no longer bothered us. Moreover, the ground was easy to work. Shellproof dugouts and shelters thirteen to twenty feet under ground had been captured in the attack, and these were impervious to the French artillery. To reach my headquarters dugout, which I shared with an attached Uhlan officer, it was necessary to crawl on all fours. The French would lay interdiction fire wherever they saw smoke; consequently we could not light any fires and during the day we were chilled to the marrow.

Ten-day reliefs were established: front line, reserve position, and rest camp alternated with one another. Thanks to the good positions and dugouts, front line losses were slight, although the French artillery increased the volume of its harassing fire from day to day. Their batteries apparently had plenty of ammunition, in contrast to ours, which were so short that we only opened fire on occasion.

I heard that the five severely wounded men we had left behind on January 29 were getting along well, and a few weeks later I was awarded the Iron Cross, First Class, for that operation. I was the first lieutenant of the regiment to get this decoration.

The next three months were spent in rectifying our lines in conjunction with our neighbors. The 120th Infantry got a bit farther forward than we on January 29. The left worked itself forward toward Cimetière, which adjoined Central on the east. Again and again sap trenches were driven forward and then interconnected. In this way the front line was moved nearer the French, and we finally reached the French wire entanglements in front of their main position.

Our work was hindered by artillery and trench mortars, the latter appearing for the first time; and many a soldier was hit in the sap trenches. The communication trenches and passages to the rear, the command posts, and the supply dumps were constantly subject to French harassing fire. When the company moved back to the rest camp, everyone heaved a sigh of relief. Usually, during such reliefs, we also had the sad duty of burying our fallen comrades. In time the reliefs became fewer, losses in the forward line increased, and the quiet forest burial ground became much more extensive.

From the beginning of May 1915, the enemy covered the foremost trenches night and day with light and medium "flying mines." The relatively quiet discharge was familiar to the experienced fighters in the Argonne. To be sure, it was much weaker than the other sounds of battle, but it was enough to wake us from the deepest sleep and drive us out of the dugouts. In daytime the missiles could be seen flying through the air, and we had ample time to duck. At night it was best to avoid the menaced area entirely. On the other hand, nobody bothered to wake up and leave the dugouts during harassing fire by the artillery.

In spite of daily casualties and the nerve-wracking strain of combat, morale remained high; everyone performing duties in an amazingly matter-of-fact manner. We even found ourselves becoming "attached" to this blood-drenched corner of the Argonne. The hardest thing was saying goodbye to buddies who were carried back severely wounded. Never to be forgotten is the soldier whose leg had been shot off by a finned bomb. On a bloody shelter-half, they carried him past us down the narrow trench at sunset. I found it difficult to express my feeling at seeing this fine young soldier leave us this way, and I could only press his hand to encourage him. But he said "Lieutenant, this is not so bad. I shall soon be back with the company even if I have to use a wooden leg." The brave lad never saw the sun rise again, for he died on the way to the hospital. This conception of duty was characteristic of the spirit of my company.

Early in May we received some prefabricated mine-gallery frames and were able to make small one-or two-man dugouts in the forward trench wall. By this arrangement we could quarter the sentry reliefs at the sentry posts. The front line was now so close to the main enemy works that the French artillery could not fire on us without endangering its own people. It transferred and increased its activity against the rear units, the supply routes, the reserve positions, command posts, and encampments.

About this time a senior first lieutenant who had not yet had field duty took over the 9th Company. The regimental commander wished to transfer me to a different company, but I declined and remained with the men whom I had previously commanded.

For ten days during the middle of May the 9th Company was attached to the 67th Infantry which was located in the middle of the Argonne near Bagatelle to the west of the 123d Grenadiers. This aggressive unit was badly depleted as a result of its many fights in the campaign. A different sort of trench warfare prevailed here. Less value was put on positions offering shelter from artillery and mortar fire. The whole battle was conducted at grenade distance from shallow depressions and from behind low sandbag walls. At Bagatelle there were few indications that the Argonne was a dense forest, for the French artillery fire had cleared the trees thoroughly, and for miles all that remained visible were stumps. While the junior commanders were making their reconnaissance prior to taking over, a brief but violent hand grenade fight broke out on a wide front; and before it was over we had suffered several casualties. This was a sample of what was in store, and we made the relief with mixed emotions.

As was our custom, we deepened the trench at once and built dugouts for ourselves. Sudden and violent bursts of French artillery and mortar fire, accompanied by hand-grenade fights all along the line, kept us from having dull moments. In the warm weather the frightful stench of corpses wafted into the position. Many French dead still lay in front of and between our positions, but we could not bury them because of the strong enemy fire.

The nights were really exciting. Hand grenade battles went on for hours along a broad front and became so confused that we never knew whether or not the enemy had broken through at some place or had worked his way behind our front line. Added to this, various enemy batteries chimed in from the flanks. This was repeated several times nightly, and we soon found it a strain on our nerves.

The platoon CP that I had inherited from my predecessors lay to the left rear of my platoon sector. At the level of the bottom of the trench—about six feet below the ground— there was a narrow vertical shaft in the forward wall. This shaft was just wide enough to permit one man to lower himself through it. Another six feet, that is, twelve feet under the surface, it opened on a horizontal tunnel the size of a large coffin. The floor was made up of cork slabs, and small niches had been cut into the wall to provide storage for rations and sundry other belongings. The walls and roof were devoid of bracing; and, while the clayey soil was holding, I knew that anyone caught there by a shell exploding near the entrance was certain to be buried alive. As soon as shells struck in the vicinity, I got out of my hole and joined my platoon. Anyway it was better to stay up forward for the nightly hand grenade fight kept us on our feet half the night.

The heat was unbearable during those days. One day Ensign Möricke, an especially fine soldier, visited me. I was down in my dugout, and we had to talk to each other through the shaft because there was not room for two in my warren. I told Möricke I was convinced that we were not safe from the damned flies even when we were twelve feet underground. Möricke said it was no wonder since the edge of the trench was simply black with them. He got a pick and started to dig there, and at the first swing the half-decayed, blackened arm of a Frenchman came to light.

We threw chloride of lime and earth on it and left the dead man in peace.

We managed to weather the ten days, and on our return to our regimental sector we were shoved back into the front line. We found that every effort had been made to render trench warfare more unpleasant, for they had added mining to an increased volume of artillery and trench mortar fire. The opposing outposts were only a few yards apart in half-covered sap trenches heavily reinforced with wire, and the night was full of lively hand grenade battles which from time to time brought the entire garrison to its feet. Each side tried to destroy the other's advance tunnels and positions and hardly a day passed without an explosion.

One day the French succeeded in cutting one of our sap trenches which had ten men of our company working in it. We got them out, but it took several hours of hard fighting and digging, for several were completely buried.

Our attempts to capture the nearby French sentry posts usually ended in considerable losses. These posts and the sections of trench leading to them were completely enclosed with barbed wire. At the slightest noise, the French in the blockhouses would sweep the obstacles with machine-gun fire. These conditions soon became exasperating and we hoped to remedy them by storming "Central."

II: Attack On Central

Following a three and a half hour artillery and trench mortar preparation, we were to seize the French strongpoints of Labordaire, Central, Cimetière, and Bagatelle. The enemy had been working on these positions since October 1914. For weeks the regiment had been making basic preparations for this attack. Close behind the front line in shellproof emplacements, medium and heavy mortars had been sited. Day and night, reserve companies brought up supplies, dismantled mortars, and ammunition through the narrow communication trenches. The French harassing fire had increased in violence, and many a carrying party had been hit. Toward the end of June and after a few days in the rest camp, the 9th Company headed back into the line. We were amazed to see large numbers of medium and heavy artillery emplaced and camouflaged in the vicinity of Binarville. It was pleasant to note that adequate ammunition seemed to be available. This time we moved up into position in the best of spirits.

The regiment prepared detailed plans for the five assault companies. During the preparation, my platoon remained in reserve about two-thirds of a mile north of "Central." Shortly before the jump-off, we were to move up behind the line of departure, follow the assault echelon closely, and keep it supplied with grenades, ammunition, and entrenching equipment.

At 0515 on June 30, the artillery opened up with everything it had including 8.3- and 12-inch mortars. The effect of the shells was unbelievable. Earthen geysers shot into the air, craters suddenly appeared before us. The strong French earthworks were smashed apart as if hit by gigantic trip hammers. Men, timbers, roots, obstacles, and sandbags flew into the air. We wondered how the enemy felt, for we had never seen such a massing of heavy fire before.

One hour before the assault, the medium and heavy mortars opened up on the blockhouses, wire entanglements, and breast-works. The French massed their artillery fire to break up the assault, but their efforts were futile. Our forward line was thinly held and was too close to the main hostile position. The French guns plowed up the dirt farther to the rear. One slug dug in a hundred yards ahead of me and tossed the mortal remains of a Frenchman who was killed in January up into the trees. I kept looking at my watch. We had fifteen minutes to go. A blue-gray pall of smoke from the bombardment obscured vision as both sides increased their volume of fire.

The communication trench assigned us was exposed to strong hostile fire, and I decided to deviate from my orders in displacing to the side for about one hundred yards. We ran for our lives across the open stretch of ground and found shelter down in the hollow; then, through the communication trench, we rushed into the front line with French standing barrages bursting all around. The storm troops were lying side by side, and across from us the last gun and mortar shells were bursting.

0845, and our assault moved forward over a wide front. French machine guns poured out their fire; the men jumped around craters, over obstacles, and into the enemy position. The assault echelon of our company was hit by machine-gun fire from the right and a few fell, but the bulk rushed on, disappearing in craters and behind embankments. My platoon followed. Each man had his load, either several spades or else sacks filled with grenades or ammunition. The French machine gun to the right was still hammering away. We passed through its field of fire and climbed over the walls on which the 9th Company stood on January 29. The position was a mass of rubble. Dead and wounded Frenchmen lay scattered through the tangle of revetments, timbers, and uprooted trees. These revetments had cost many a Frenchman his life.

To the right and in front of us hand-grenade fights were in progress, and from rearward positions French machine guns swept the battlefield in all directions and forced us to take cover. The sun was hot. Stopped over, we moved off to the left and then, hard on the heels of our company's assault echelon, pressed on to a communications trench leading to the second position.

Our artillery had shifted its fire to the second French line (Central II), located 160 yards to the south, which was to be captured on July 1 only after renewed artillery and mortar bombardment. The assault echelons of the regiment not engaged in mopping up Central I, were pushing on toward Central II.

About thirty yards ahead of us a violent handgrenade fight raged, and beyond that we could see the outlines of Central II some ninety yards distant. French machine-gun fire made it impossible to move outside the communication trench, and our own assault group up forward seemed to be stalled. Its young leader, Ensign Möricke, was lying in the trench severely wounded with a bullet in his pelvis! I wanted to carry him back, but he said that we should not worry about him. Stretcher bearers took charge. One last hand-clasp and I assumed the command up front. The Ensign died the next day in the hospital.

We engaged the garrison of Central II. Our own artillery had ceased firing. A few salvos of hand grenades followed by a charge, and we were in Central II. Part of the garrison ran down the trench, others fled across the open fields, and the rest surrendered. While part of the outfit worked to widen the trench, the bulk of the assault force kept pressing south. We moved through a ten-foot-deep communication trench and had the luck to surprise and capture a French battalion commander with his entire staff. About one hundred yards down the line the trench opened on a large clearing. In front of us the terrain dropped sharply down to the valley at Vienne-le-Château, which was obscured from view by the woods. We lost contact on both flanks and to the right. On the edge of the forest some two hundred yards away we saw a considerable number of the enemy. We opened fire and after a sharp fire fight they withdrew into the woods. While this was going on I established contact on my left with elements of the 1st Battalion who had advanced this far; and I also reorganized my command, which by this time included elements of all units of the 3d Battalion, and deployed it in a defensive position some 350 yards south of Central II. Because of our exposed right flank and since we still heard the sounds of bitter fighting behind us, I felt it inadvisable to continue the advances to the south. Also the memory of January 29, when I went so far out front that I lost all support, was still fresh in my mind.

A reconnaissance detachment reported that the unit on our right had been unable to clear Central I. This meant we must prepare blocking positions to protect our newly-won positions against attacks from the west. To bolster the defense I put my best veterans in line and was glad of it, for during the next few hours the French launched a series of violent counter-attacks to regain their lost positions. I kept the battalion commander informed of developments.

On the left some companies of the 1st Battalion had advanced down the valley as far as Houyette gorge. Combat outposts reported strong enemy forces in the woods on the slope 330 yards ahead. I discussed the situation with Captain Ullerich, the commander of the 1st Battalion, and he decided to have the 1st Battalion dig in on the left of the 9th Company.

We wasted no time in getting to work. I kept one platoon in reserve and used it to bring up ammunition and hand grenades and for work on the flank position in Central II. French reconnaissance detachments were probing our front, but we drove them off without trouble.

Digging was easy, and in a short time our trench was better than three feet deep. Since the start of the attack the French artillery had been rather quiet, but now it opened up on Central II with every weapon at its disposal. The French apparently believed we occupied the place in force, for their ammunition expenditure was very heavy. The net result was to smash up their old position and cut our communications to the rear. Supplies ran a gauntlet of fire, and our lone wire was soon knocked out. We did succeed in emplacing one heavy machine-gun platoon in the company sector.

By evening our trench was five feet deep, and the French artillery was still falling behind us. Suddenly bugle calls rang out in the woods; and the enemy, in his usual massed formation, rushed us from the woods an eighth of a mile away. However, our fire soon drove him to earth. There was a slight fold in the ground, and as a result we picked up targets from our trench only when they came to within ninety yards of our position. Perhaps a position farther back near Central II would have been better. We certainly would have had a good field of fire; but, on the other hand, the French artillery would have punished us severely. The French attacked with vigor, and hand-grenade fights continued all over the place even after it became quite dark. Our supply of grenades was limited and we did most of our fighting with the rifle and heavy machine gun. The night was black and the smoke from bursting grenades reduced the effectiveness of our rocket flares. Grenades burst all around us, for the enemy was within fifty yards of our position. The fight waxed and waned all night, and we beat off all attacks. At daybreak we made out a sandbag wall some fifty yards away and all sounds indicated that the enemy were busily engaged in digging in behind their improvised shelter. The French infantry kept us on the go during the night and their artillery then took over the morning shift. Fortunately the bulk of artillery fire landed in Central I and II with but a small fraction striking close to our position, and a hit in the front line was a great rarity. So we felt comparatively secure and did not envy the carrying parties who had to move rations and other supplies up through the shell-plastered ammunition trenches.

We spent the succeeding days in improving the position. The trench was soon six feet deep; we constructed small wood-lined dugouts, installed armored shields, and sandbagged firing points. The front-line losses from artillery fire were few, but we lost men daily in the shell-torn communication trenches leading to the rear.

The strong artillery concentrated for the June 30 offensive moved to another front, and our weak organic artillery lacked sufficient ammunition to give us worthwhile support. However, an artillery observer was always up front, and we in the infantry appreciated this very much.

During the early days of July the enemy began to wreck our trenches daily with flying-mines fire delivered from positions that permitted a certain degree of enfilade fire. These flying-mine projectors are of simple construction and provided but little lateral dispersion; consequently he obtained a high percentage of direct hits, and it was not always possible to evacuate the dangerous localities in time to avoid losses. Our casualties from these mines were considerable, several men being killed by the blast effect of the 220-pound bomb alone.

During July I began a five weeks' assignment as substitute of the commander of the 10th Company in a sector where relief was furnished by the 4th and 5th Companies. We company commanders worked on a unified plan for construction of shell-proof multi-entrance dugouts twenty-six feet under ground. This work went on day and night with several parties working on one dugout from different directions. The officers pitched in, and we found that sharing the work helped morale.

Frequently an entire position was flattened by artillery inside of an hour. When this happened we saw our lightly timbered dugouts collapse like cardboard boxes. Fortunately the French bombardment followed a rigid plan. Usually they began on the left and traversed right. To remain under heavy fire proved too expensive; so whenever it started I evacuated the trench and waited until they moved their fire laterally or lifted it to our rear area. If French infantry had followed their artillery and attacked us, then we would have thrown them out with a counterattack. This did not bother us because, man to man, we felt ourselves to be superior.

The Central I operation repeated itself, and we began to drive short saps and mine tunnels toward the enemy position. Early in August my company relieved the 12th Company in the Martin sector. This outfit needed relief following heavy losses sustained the day before in an attack made after the blowing of a mine tunnel. The relief was made at dawn without incident, but we were hardly in the position before the French artillery opened up, and we spent some tense moments flat on our faces with enemy dead lying all around us. When the fire slackened and died down we got our spades working and began deepening the position. Once the trench was eight feet deep and had had numerous small dugouts let into the forward wall, we stopped worrying about the French field artillery. Anyway I wanted to come out with the same number with which I went in.

The heavy spade work paid; and two days later, in spite of much harassing fire, we left the position with only minor casualties. I turned the company over and took my first leave of the war.

Observations: On June 30, in order to deceive the enemy as to the time of attack, there were numerous breaks in the three-and-a-half hour artillery and mortar preparation. In spite of very heavy fire the hostile position was not completely destroyed, and a few machine-gun nests still put up resistance during the attack.

The great offensive power of the German infantry was again evident. It did not stop on the initial objective but went ahead and seized the next French positions. The speed of attack was such that a battalion commander and his staff were surprised and taken prisoner. The switch from successful offense to defense was made rapidly; and since the French were thoroughly familiar with their old positions, we avoided using them. Detailing of carrying parties for ammunition and tool supply was excellent foresight, for French retaliatory fire prevented the resupply of the assault echelon for several hours and also interrupted telephone communications.

On July 1, rifle and machine-gun fire played a dominant role in the repulse of French counterattacks launched against us from the nearby woods.

Before daybreak and using a sandbag wall as cover, the French infantry dug in some fifty yards in front of our line. Apparently a part of these sandbags had been taken along during their attack or had been brought up by rear units after the attack bogged down.

In the weeks following our attack, when strong enemy artillery fire was directed at the position, sections were evacuated frequently in order to reduce losses. Current infantry regulations allowed a company commander on the defense to execute local retirements when subjected to heavy harassing fire.

III: Attack of September 8, 1915

On my return from leave, I got the 4th Company which a few days later was to attack on the right flank of the regiment. I took over the company in a reserve position in the Charlotte valley. After a personal reconnaissance of the assembly area and the terrain to be attacked, I held a few rehearsals in the old valley positions and in this manner prepared the company to approach the tough job that lay ahead with complete confidence. I regretted that my command lasted only a few days, but I was too junior to be a permanent company commander.

Before dawn on September 5 my platoon moved up through the communication trenches in a confident mood. The position we took over from a company of the 123d Grenadiers was being undermined by the French. At various places we could clearly hear the incessant work of the hostile tunneling squads. We hoped that the enemy would not stop burrowing before the start of the attack. We preferred honest man-to-man fighting to being blown into the air. Three long days passed during which the moles beneath worked without interruption.

At 0800 on September 8 our artillery and mortars opened on the hostile positions located forty to seventy yards in front of us, and its volume of fire was not inferior to the preparation used against "Central." The French artillery replied with all guns at its disposal; and we crouched in our small, flimsy dugouts and weathered the bombardment which tore up the landscape and showered our trench with a continuous rain of dirt, splinters, stones, and branches. The fire even uprooted some century-old oak trees. All during the shelling we were unable to locate the French miners working underneath us. Was their work finished?

From time to time I toured the company sector to check and see how my men fared, and on more than one occasion I was bowled over by the blast of a shell exploding nearby. I took a look over at the enemy and the sight that met my eyes must have been similar to the one he saw when he looked at us. The air was filled with flying debris, a gray-blue pall hung over the rearward parts of his installations.

This softening-up process took three hours and finally the watch hands crawled to 1045.

Crouching, the three company assault teams moved to their jump-off position. No synchronized watches since the attack jumped off immediately following the last shell burst at 1100. The engineer squads and the ammunition and matériel-carrying parties arrived, and I pointed out the squad objectives to each squad leader. These objectives were some 225 yards beyond the enemy line, and I impressed upon them that it was necessary to keep to their objective and that elements of the company following us in second line would take care of any outside local resistance.

The activities after the successful attack, the consolidation of the gains, the establishment of contact, and the blocking off of sectors were thoroughly discussed.

Meanwhile, under a terrific concentration of fire, all types and calibers of shell were demolishing the hostile position. It was hard to imagine that anything remained alive when the gunners got through. Thirty seconds more! The riflemen crouched in the craters ready to go. Ten seconds more! The last shells struck close ahead of us; and, before the smoke cleared, our three assault teams rose noiselessly from their trenches and dashed for their objective on a front of some 280 yards. The attack moved through the smoke and noise of the battlefield with the same precision shown in the rehearsals of the past few days. A wonderful picture!

Our men ignored the crowds of fear-crazed French soldiers climbing out of the nearest positions with hands raised high and showed the prisoners the way toward our line of departure. The assault squads rushed toward their objectives, and the units following in the second line under the company first sergeant took care of the prisoners.

I joined the assault team on the right. We rushed forward past enemy trenches and, in a few seconds, reached our objective. Engineers, entrenching squads, and hand-grenade squads followed close on our heels and, up to this time, no one had been hit. Our advance was without the usual accompaniment of cheers and shouts and as a result of this silence we took the French rear positions completely by surprise. These people were convinced that the jig was up and they surrendered without offering any resistance. A machine gun opened up and forced us to take cover. We moved to the left down the trench and established contact with the center assault team and a few minutes later we were in contact with our left unit and with the adjacent company (2d) as well.

We started work consolidating our positions for defense and in a very short time the trenches leading toward the enemy were blocked off with sandbags and ammunition and grenade dumps had been established. The French artillery opened up on the area in rear of us in such volume that we were completely cut off from our jump-off position. French machine guns prevented any movement outside the newly won position and we had no chance of immediate resupply. The French infantry counterattacked but even though we had a field of fire one hundred yards in width, we had little difficulty in stopping them. Within the positions, violent hand-grenade fights centered around the blocked-off trenches, but there, as elsewhere, the enemy made no gains. The terrain sloped gently toward the enemy, and our hand grenades reached farther than his.

During the attack five men in one of the assault teams were put out of action by an improperly thrown grenade. French fire after we reached our objective caused a total of three dead and fifteen wounded for the company. Supplies of all classes were our next problem. Ammunition, matériel, and provisions had to be brought across open terrain which was constantly swept by French machine guns and artillery fire. A communication trench back to the line of departure had to be dug and we had to establish contact with the unit on our right.

Upon my suggestion, the battalion commander decided to have eighty men from the reserve dig a hundred-yard trench from our position back to the jump-off. This work was placed under my direction. Our work was fifty yards from the French so I ordered my matériel squads to bring up a large number of sandbags and steel shields. The French, on June 30, taught us a good lesson.

We started work at 2200 although the enemy was still restless and excited and kept up an almost continuous fire illuminated by rocket flares. To get the job done in a single night meant getting under way without delay. To start with, I ordered a sixteen-inch sandbag wall to be pushed out from each end of the still nonexisting trench. To build this wall was hell and we formed an endless human chain with each man flat on his back and in this way got the bags out to the wall-builder. Hostile small-arms had no effect on the men behind the bags and in no time the wall stuck out into open space for a distance of fifty feet on each end. Then the supply of sandbags gave out, leaving a gap 230 feet wide. I closed this gap by having men take their armored shields into the gap and form a skirmish line. As soon as the individual soldier got to his place, he seized the steel shield and began to dig in behind it. Rifle and hand grenades were at hand. This whole movement was accomplished without noise although the enemy fired numerous flares and used all his infantry weapons. The only thing that reached us was rifle fire and that could not penetrate our armored shields. Even so, the skirmish line was none too comfortable. When the morning of September 9 broke we had a six-foot communications trench leading back to our old position.

Just as I hit the hay after a hard day's work, the battalion commander and, hard on his heels, the regimental commander arrived to inspect the new positions. They were pleased with the success of the 9th and 2d Companies. The assigned objective had been seized. Some officers, 140 enlisted men, sixteen trench mortars, two machine guns, two drilling machines, and one electric motor were in our hands. The joy of the 4th Company over the success was dampened by the death of Lieutenant Stöwe, our liaison officer with the 123d Grenadiers and who had leave orders in his pocket.

Shortly after the attack I gave up the 4th Company again and, for a few weeks, took over the 2d Company. I was none too happy for the 4th and I got along extremely well. In command of the 2d Company, I spent some time in Crown Prince Fort, a shellproof shelter and blocking position 160 yards behind the front line. While there I was promoted first lieutenant and transferred to a mountain outfit which was to be activated in Münsingen. It was hard to leave a fighting regiment in whose ranks I had fought through so many a hard day and to say goodbye to the many gallant soldiers and to the blood-soaked, hotly contested soil of the Argonne. The Champagne battle was at its peak when toward the end of September I left the Binarville forest.

Observations: With the newly acquired company the attack for September 9 was thoroughly rehearsed. The three assault squads were to advance as soon as the artillery preparation ceased, traverse the nearby enemy position without shooting, and capture the assigned objectives some 220 yards away. Mopping up operations were assigned the company units following in second and third lines.

Contrary to my orders, one assault team used hand grenades during the advance and wounded five of our men. (These were the only losses sustained in the attack proper. In principle: Do not throw hand grenades during a charge, for our own people will run into them. The surprise achieved was complete. We were past the enemy forward position before he could grab a rifle, and our appearance at the entrance to his rear dugouts must have been regarded as Mephistophelian. The result was a relatively large number of prisoners.

After the attack, we shifted swiftly to defense, this time by using the positions at hand, and we had no trouble in repulsing their counterattacks. Again, after the assaults, the company communications to the rear were interrupted for hours by artillery and machine-gun fire. The use of sandbags and armored shields greatly simplified the job of making contact with the unit on our right.

Chapter 6

RAIDS IN THE "PINETREE KNOB" SECTOR, HIGH VOSGES

I: The New Unit

Early in October the Württemberg Mountain Battalion (six rifle companies and six mountain machine-gun platoons) was activated near Münsingen and placed under the command of Major Sprösser. I had the 3d Company composed of two hundred young veterans drawn from all branches of the service. We had a few short weeks to train and produce an efficient mountain unit. The variety of uniforms lent color to our formations and morale was high from the first day. Officers and men put everything into the training program and our rigorous régime soon produced results. The new mountain uniform which was issued later was most becoming.

Toward the end of November Major Sprösser, a martinet, held his final inspection and, early in December, we were transferred to the Aarlberg for ski training.

The 2d Company was quartered in the St. Christopher Hospice near the Aarlberg Pass. From early morning until dark, with and without packs, we practiced skiing on the steep slopes. During the evenings, we sat around the improvised day room and listened to the songs, mostly mountain ditties, played by our company band under the direction of Father Hügel. This certainly was a change from the Argonne of a few months ago. This off-duty contact improved my acquaintance with my men and tightened the ties between us.

We enjoyed the Austrian ration which included cigarettes and wine, but we felt that we earned our keep. Christmas was celebrated in the best of spirits.

This wonderful living ended all too soon and four days after Christmas we boarded a troop train heading west instead of for the Italian front as we had hoped. During a rainy, howling New Year's Eve, we took over the South Hilsen Ridge sector from a Bavarian Landwehr outfit.

Our new sector was ten thousand yards long with an elevation differential of five hundred feet between right and left. Strong wire entanglements, one electrically charged, and other obstacles lined our front. Of course, it was impossible to provide a continuous garrison for this frontage and we developed and strengthened certain commanding points along the line. These strongpoints were miniature forts, organized for all-around defense and stocked with ample supplies of ammunition, rations, and water. I put my Argonne experience to good use and made certain that each dugout had two exits as well as strong overhead cover.

The French positions were not within grenade range as in the Argonne, but only approached to within a few hundred yards of us along our right flank and in the center, with the rest farther away along the edge of a thickly wooded area. During the spring and summer we learned to know the various positions: the Little Southern, the Whip, the Picklehead, and the Little Meadow. Also, during this period we spent much time in training our numerous officer candidates.

In September, we took over the open position on the northern slope of Hilsen Ridge. The French were close by and we received heavy artillery and trenchmortar fire as part of our daily diet.

II: Raids in the "Pinetree Knob" Sector

Early in October 1916, several units, among them the 2d Company, were ordered to prepare plans for raiding the enemy to take prisoners. My Argonne experience showed me that this type of work was dangerous, difficult to organize, and usually resulted in a high casualty list and, for this reason, I held my boys in check as far as raids were concerned. Once ordered, I pitched in and started planning. To begin with, in order to determine the feasibility of working into the enemy position, I went out on reconnaissance, accompanied by Sergeants Büttler and Kollmar. We crept and crawled through the tall and fairly dense fir forest toward a French sentry post, established at the upper end of a forest path leading up toward the enemy. The path was overgrown with tall grass and woods and we were most careful in crossing it some fifty yards from the enemy. Once across we slipped into a ditch and wormed our way forward. Cutting through the mass of barbed wire with cutters was a back-breaking job requiring a maximum of caution. Night began to fall and we heard but could not see the French sentry move about his post. It was slow work getting through the wire, especially since we could only cut the lowest strand of wire. Eventually we reached the middle of the entanglement. At this point the French sentry exhibited a certain amount of restlessness and cleared his throat and coughed several times. Was he afraid or had he heard us? If he tossed a hand grenade into our ditch, it was curtains for the three of us. To make matters worse, we could not move let alone defend ourselves. We held our breath and let the tense moments tick by. As soon as the sentry quieted down, I began to withdraw. By this time it was totally dark. In crawling back, we snapped a few twigs and this inadvertent act received immediate recognition. The enemy alerted his entire position and for minutes sprayed the entire landscape between positions with all varieties of small-arms fire. We hugged the ground and let them pass over us. When silence fell, we resumed our journey and returned without suffering any casualties. Our reconnaissance proved conclusively the difficulty of raiding in those wooded sectors.

Next day, I proceeded to examine the possibilities of getting into a hostile position called Pinetree Knob, and found the situation more favorable. Under cover of darkness, the position entanglements could be reached silently by moving up a grass-covered glade. However, the barriers consisted of three separate belts of wire requiring hours to cut. Only five hundred feet separated our trenches from those of the enemy. It took several days and nights of careful reconnaissance before we were able to determine the exact location of two sentry posts on Pinetree Knob. One was located in the center of a glade in a concealed sentry box; the other was two hundred feet to the left on a rock ledge from which it was easy to cover the surrounding terrain with fire as well as with visual observation. Only on rare occasions did we receive machine-gun fire from this part of the sector.

Any operation in this direction moving over grassy terrain totally devoid of cover required a moonless night. During the next few days and nights we studied the details of the approaches into the Pinetree Knob position and observed the personal habits of the garrisons of both outposts. In doing so, we carefully avoided drawing our opponent's attention to the impending expedition.

I based my plans on the results obtained through our reconnaissance. This time I did not intend to sneak into the positions proper; I proposed to negotiate the wire field midway between the two posts, get in this trench and then strike them in flank or preferably in the rear. The raid required a force of twenty men, for we had to split up on arrival in the hostile trench. Also I had to plan to get my raiding

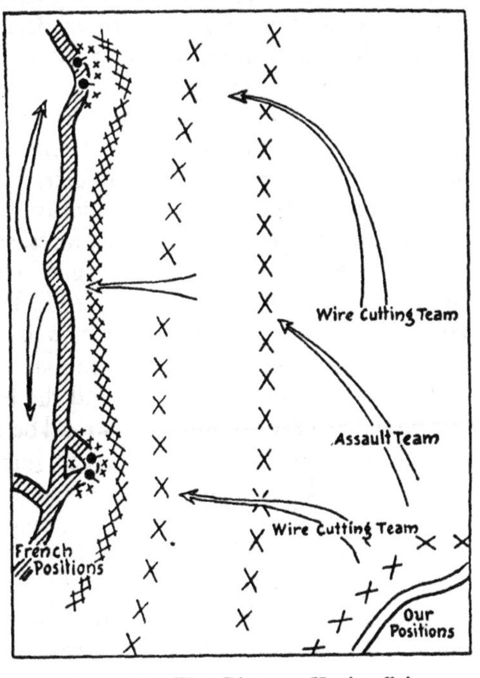

Sketch 12: The Pinetree Knob affair.

parties out and I had to allow for a possible attack by the local trench garrison. A wire-cutting party was to be placed opposite each enemy outpost. They were to crawl up to the edge of the wire and remain there doing nothing until the raiders started cleaning out the trench with pistols and hand grenades or until signalled from within the captured outpost positions. Not until one or other happened were they to start cutting the wire and make a path for the raiders' escape.

I discussed the raid with the subordinate leaders, using sketches and pointing out the terrain from the trench. The various detachments started their preparations by holding rehearsals close behind our positions. October 4 was a cold and nasty day. A strong northwest wind drove clouds through our 3500 foot position. Toward evening the wind changed into a storm, and a rain of cloudburst proportions beat down. This was weather I had been praying for. By that time the French sentries had their heads deep in their coat collars and had taken refuge in the most sheltered corner of their sentry hole, thus reducing their efficiency as guards. In addition, the wind would drown out most of the noise occasioned by our approach and wire cutting. I told Major Sprösser that the night had arrived and he said, "Go ahead."

Three hours before midnight, in a pitch black, stormy and rainy night, I left our positions with my three detachments and crawled slowly toward the hostile position. Soon the wire-cutting detachments under Staff Sergeant Kollmar and Lance Corporal Stetter left us and moved off to the right and to the left. Lieutenant Schafferdt, Staff Sergeant Pfeiffer and I went with the assault detachment and crawled behind our wire-cutters. The other twenty soldiers followed in single file at three-pace intervals. We crawled noiselessly toward the enemy. The wind howled and whipped rain into our faces, and soon soaked us to the skin. We listened anxiously into the night. Single shots rang out here and there and an occasional rocket flare flickered in the darkness, but the enemy remained quiet. The night was so dark that the silhouettes of the surrounding rocks were indistinguishable at more than sixteen feet.

We reached the first obstacle and the hard work began. One of our trio wrapped a rag around every strand before using the cutters. Another took the tension off the wires, and the third slowly cut the wire through. The ends of the cut wire were bent back carefully to prevent betrayal by the noise which would result were they allowed to fly back unimpeded. Every move had been tested beforehand.

We stopped work occasionally and listened intently into the night before starting our tiresome job again. In this manner, inch by inch, we cut our way through the high, wide and very finely laced French wire. We had to content ourselves with cutting a path through the lower strands only.

Those were hours of strenuous work! Occasionally, a wire cracked and we stopped work and strained our ears to listen into the night. By midnight we had cut our way through the second wire belt and we were a hundred feet from the enemy trench. Unfortunately the rain and the storm abated somewhat, and it had become a bit lighter. Ahead of us lay high and continuous *chevaux-de-frise*. Each separate framework was long and heavy and the innumerable wires were too heavy for our light wire cutters. We crawled a few yards to the right and tried to separate two of the *chevaux-de-frise*. This effort merely made a lot of noise, which sounded to us like a thunderclap. If the hostile sentries, now a hundred or so feet away, failed to hear us, they were surely asleep.

The next few minutes were not pleasant, but all remained quiet on the southern front. I gave up trying to separate the *chevaux-de-frise* which were too firmly anchored and, after a brief search, we found a shell crater which gave us an opening. We wormed our way through and covered the few yards between us and the enemy position.

Another shower started. The three of us were between the wire and the enemy trench where the water trickled down the trench bottom over some stone steps and on into the valley. Cautiously, the leading men of the assault detachment squeezed under the *chevaux-de-frise*. The remainder were farther back in the first and second wire belts. Suddenly we heard footsteps coming down the trench from our left. Several Frenchmen approached us coming downhill in the trench and their slow and even steps resounded in the night. They were unaware of our presence. I estimated their strength at three or four men. What were we to do? Jump them or let them pass? The chances of doing all this without raising a ruckus were remote. It would be a man for man fight. Our own assault detachment could not take part for it was still out in the wire. We could have overpowered the trench patrol, but then the trench garrison would have gone into action and covered the barriers with fire. Our return would have cost us dearly and, under such conditions, we would have had little luck in bringing back a prisoner. I quickly weighed the pros and cons and decided to let the enemy pass unmolested.

My two companions, Schafferdt and Pfeiffer were informed and we took complete cover at the edge of the hostile trench for, above all, we had to hide our hands and faces. The *chevaux-de-frise* interfered with our crawling back. We would have been detected had the French patrol been on the job. In case they were, we got ready to jump them. With our dispositions made, we lay and waited. Their footsteps were regular and they conversed softly. Anxious seconds crawled by. Without hesitation, the French trench patrol came abreast of us and went on. While the sound of footsteps died away, we heaved a sigh of relief and waited a few minutes to see if they would return. Then, one after the other, we dropped into the trench. The rain had stopped and only the wind whistled over the bare slope. As the wary men entered the trench, bits of earth and rock broke loose from the trench wall and tumbled noisily down onto the stone steps. Again anxious minutes passed. Finally the whole assault detachment was in the trench.

We divided and Lieutenant Schafferdt, with ten men, went down the slope while Staff Sergeant Schropp and his ten went in the opposite direction. I went with Schropp. We felt our way carefully up the steep trench. Only a few steps separated us from our objective, the sentry post on the rock ledge. We wondered if the enemy had noticed anything. We stopped and listened. Suddenly over on the left something smacked into the barrier, followed immediately by an explosion on trench parapet on the right. Hand grenades exploded with a roar. The leading man of the assault detachment reeled back, and the whole detachment became jammed in the trench. The next hand grenade salvo struck among it. It was a question of attacking immediately or surrendering. "Let 'em have it!" We rushed the enemy and managed to pass under his hand-grenade fire. Stierle, my groom, who had come up forward for this party only, was hit on the larynx by a Frenchman, and Sergeant Nothacker dispatched the man with his pistol. A short time afterward, two other men of the sentry detachment were overpowered. One Frenchman managed to escape to the rear.

With our flashlights, we made a hurried search for dugout entrances. We found one hole that was empty, but a second one was full of Frenchmen. With my pistol in my right hand and flashlight in my left, I crawled into the twenty-inch opening followed by Sergeant Quandte. Seven fully-armed Frenchmen sat along the wall, but they threw down their arms after a brief argument. The safest course was to take care of these lads with a grenade or two, but this was contrary to our orders, which specified that prisoners were to be brought back.

Lieutenant Schafferdt reported two prisoners with no losses in his unit. While we were occupied with the job at hand, the wire-cutting detachments had been working like beavers and the paths through the wire were ready.

Since the *coup* had accomplished its purpose, I gave the order to withdraw. We had to break away before French reserves got into action. Without further annoyance from the enemy, we regained our position with a bag of eleven prisoners. Particularly pleasing was the fact that we suffered no real casualties. Lance Corporal Stierle had a slight scratch from a hand-grenade fragment. Recognition by our superior officers for this fine operation was not long in coming.

Unfortunately the next day brought retribution, for a French sniper picked off Staff Sergeant Kollmar in a quiet sector of the company trench. This lamented loss deepened our joy over the successful Pinetree Knob affair.

After this the days in the "Open Position" were numbered. The Supreme Army Command had other work for the Württemberg Mountain Battalion. Toward the end of October we moved east.

Part Three

OPEN WARFARE IN RUMANIA AND THE CARPATHIANS 1917

★ ★ ★ ★ ★ ★ ★ ★ ★ ★ ★ ★ ★ ★ ★

Chapter 7

FROM SKURDUK PASS TO VIDRA

I: Occupation of Hill 1794

In August 1916, the Central Powers' front was subjected to powerful assaults by the Entente armies. On the Somme enormous English and French forces struggled for a decision. The fire flared up anew on the blood-soaked fields around Verdun. In the east, the front was still shaken from the effects of the Brussilov offensive which had cost our Austrian allies half a million men. In Macedonia a large Allied army under General Sarrail, stood ready to attack. And on the Italian front, the sixth battle of Isonzo had ended with the loss of the Görz (Gorizia) bridgehead and the city of Görz. Here too the enemy prepared new offensives.

At this point the Rumanians marched onto the stage as new enemies. They believed that their entry into the war would result in a quick Entente victory. As reward they expected much from their allies. On August 27, 1916, Rumania declared war on the Central Powers and half a million Rumanian soldiers crossed the frontier pass and moved into the Siebenbürgen district. When toward the end of October, the Württemberg Mountain Battalion arrived in the Siebenbürgen district, far-reaching victories had already been won in the Dobrudja, at Hermannstad and at Kronstadt, and the Rumanians had been thrown back across their frontiers, but the decisive battle remained to be fought. The Russians reinforced the Rumanian Army, which a few weeks before had crossed the frontier with the brightest of hopes, but had been forced back across that frontier.

The Württemberg Mountain Battalion detrained at Puy on the wrecked railway line which ran to Petrosceny. The hard march toward Petrosceny continued over churned-up roads which were blocked by columns of every description. The following expedients proved effective for getting ahead. The leading squads of the Company marched with fixed bayonets. They cleared a way through the confused traffic which time and again blocked the way. The company's vehicles were accompanied by riflemen. The men took hold whenever the horsepower threatened to fail. With this arrangement, the troops moved slowly but steadily forward. We met Rumanian prisoners in their high, pointed caps.

Shortly before midnight the Company arrived in Petrosceny and slept for a few hours on the bare floor of a schoolhouse. Our feet were burning from the long march. Nevertheless, before dawn, the 2d and 5th Companies climbed into trucks and traveled through Lupeny toward the threatened mountain front.

A few days before, the drive of the 11th Bavarian Division through the Vulcan and Skurduk Passes had failed. In bitter fighting for the pass exits, parts of the infantry and the artillery were thrown back and badly scattered. At present, the Schmettow Cavalry Corps was in possession of a ridge running along the border. Had the Rumanians continued their attack, it would have been difficult for our weak forces to hold them in check.

After a truck ride of several hours, we detrucked in Hobicauricany. Here the cavalry brigade to which we had been attached started us toward the border range in the direction of Hill 1794 (5920). We climbed over a narrow footpath and our packs with their four days' uncooked rations weighed heavily on our shoulders. We had neither pack animals nor winter mountain equipment, and all officers carried their own packs. We climbed the steep slopes for hours. We met a few men as well as an officer from a Bavarian unit which had fought on the other side of the mountain. Their nerves appeared to be pretty well frayed. According to their stories, they had had an extremely tough time in a battle in the fog; and a majority of their comrades had been killed in close fighting with the Rumanians. For days and without food these few survivors wandered through the mountain forests and had finally found their way across the frontier range. They described the Rumanians as wild and dangerous adversaries.

Late in the afternoon, we reached an altitude of 3960 feet and located the sector CP. While the various companies cooked supper, Captain Gössler (commander of the 5th Company) and I were given the situation and ordered to continue the march as rapidly as possible, reach Hill 1794 (5920) that same evening, occupy the positions on top of it and reconnoiter southward through Muncelul and Prislop. The latest reports from the reconnaissance troop which had penetrated south of Muncelul were two days old. A telephone station and led horses were supposed to be on Height 1794 (5920). Contact did not exist with the units on the right and left.

It began to rain as we started to climb without benefit of a guide. The rain grew heavier as night began to fall and it was soon pitch black. The cold rain turned into a cloudburst and soaked us to the skin. Further progress on the steep and rocky slope was impossible, and we bivouacked on either side of the mule path at an altitude of about 4950 feet. In our soaked condition it was impossible to lie down and as it was still raining, all attempts to kindle a fire of dwarf pine failed. We crouched close together, wrapped in blankets and shelter halves and shivered from the cold. As soon as the rain slackened, we again attempted to build a fire, but the wet pine branches only smoked and gave out no heat. Slowly the minutes of that terrible night crept by. After midnight the rain ceased, but in its stead an icy wind made it impossible for us to relax in our wet clothes. Freezing, we stomped our feet around the smoking fire. Finally it became light enough to continue the climb toward Height 1794 (5920), and soon we reached the snow line.

When we reached the summit, our clothes and the pack were frozen to our backs. It was below freezing and an icy wind was sweeping the snow-covered summit of Height 1794 (5920). Our positions were not to be found. A small hole in the ground, barely capable of holding ten men, sheltered the telephone squad. Over on the right were some fifty shivering led horses. Shortly after our arrival a blizzard enveloped the elevated region and reduced visibility to a few yards.

Captain Gössler described the situation to the sector commander and tried to have the two companies withdrawn. However, all representations of the experienced alpinist were in vain, even though the surgeon also warned that a continued stay in the snowstorm in wet clothes, without shelter, without fire, and without warm food, would result in many sick and much frostbite within the next few hours. We were threatened with court-martial proceedings if we yielded one foot of ground.

To ascertain the whereabouts of the missing troop, Staff Sergeant Büttler was sent in the direction of Stersura via Muncelul; the mountain troops pitched tents in the snow. We did not succeed in making fire. Numerous cases of high fever and vomiting were reported, but renewed representations to sector were without effect. A horrible night began. The cold became more biting and soon the men could not stay in their tents and, as on the previous night, tried to keep warm by moving about. A long, long winter night! When day broke the doctor had to evacuate forty men to the hospital. I was on orders from Captain Gössler to the sector commander to give a personal description of conditions on the summit, and I at least succeeded in having our request forwarded for immediate disposition. When I returned to Hill 1794 (5920), Captain Gössler had decided to move off with the remainder of the companies, come what may; ninety per cent were under medical treatment because of frostbite and cold symptoms. The weather cleared at noon just as we were being relieved by fresh troops equipped with pack animals, wood and other items of equipment. Meanwhile the reconnaissance troop had been discovered by the Büttler scout squad on one of the southern spurs of the mountain. There at an elevation of thirty-six hundred feet, bearable temperatures prevailed. There was no trace of the Rumanians.

After three days the company was back in good shape. Under considerably more favorable weather conditions and with better equipment, we climbed the Muncelul. After a bivouac at 5940 feet, we moved forward toward Stersura, a foothill of the Vulcan mountains which drops perpendicularly to the northeast and north. The company sent outposts about eleven hundred yards north of Stersura. While they were digging a hedgehog on a wooded knoll, secured by three sentries, things got lively on the Stersura. Rumanians in about battalion strength were dug in across the way in several superimposed positions.

During the ensuing days encounters with a weak enemy resulted in no losses on our side. We lived in tents near our positions; pack animals brought provisions daily from the valley on the other side of the mountain ridge; telephone communication connected us with the Sprösser group and with our sentries. Over on the right was Arkanului. On its steep southeast slopes we could see units of the 11th Division's artillery which had been abandoned there. About a mile and a third east of us, on the next ridge, were other units of the Württemberg Mountain Battalion.

Fog covered the plain far below us and broke like ocean waves against the sunlit peaks of the Transylvanian Alps. A wonderful sight!

Observations: The occupation of Hill 1794 (5920) showed how high mountain weather can influence the efficiency and resistance of the troops, especially when the equipment is not suitable and complete, and supply fails. On the other hand, we saw what the soldier can endure in presence of the enemy. Under certain circumstances dry wood or charcoal must be furnished for troops living at an elevation of six thousand feet. A few days later, on the southern slopes of the Vulcan mountains, we were heating our tents with small charcoal fires built in suspended tin cans.

II: Attack on the Lesului

In November the Rumanians were prepared for a German thrust from Kronstadt in the direction of Bucharest and they had the bulk of their reserves concentrated in the area north of Ploësti. They were blissfully ignorant that General Kühne was forming a new attack group in the

Vulcan-Skurduk area for the purpose of forcing an entry into Wallachia and advancing on Bucharest from the west.

Early in November units from our battalion, on the right of the new group, seized the line of heights running from Prislop through Cepilul to Gruba Mare. This operation was designed to provide protection for the debouchment of our main forces from the mountains. We had to fight hard and once we were in possession we worked hard to put our newly-won gains in condition to meet the inevitable counterattacks. The Rumanian fought well but all his counterattacks were beaten back and he began to wire himself in on the Stersura. On November 10 my company, less one platoon which was left behind on a security detail, was moved to Gruba Mare to participate in the attack of the Kühne Group. The attack was scheduled for the eleventh and our Battalion mission was to seize the Lesului, a commanding peak some four thousand feet in elevation whose southern slopes formed part of the Wallachian frontier. The Rumanians had fortified this peak to the best of their ability and we could see several hostile positions located one behind the other in the saddle between Gruba Mare and Lesului. Our battalion mustered four and a half rifle companies (among them the 2d) for this attack and a mountain battery of artillery was attached in direct support. Gössler's detachment was to make a frontal attack while Lieb's was enveloping the hostile position from the east. Lieb had two and a half companies for his part of the job. The frontal attack was to jump off only after the enveloping force had been committed to action.

The 2d Company was reinforced with a machine-gun platoon and dawn of November 11 found us on the right of the position a scant two hundred yards from the Rumanians. We were ready to go. On our way into the assembly area we ran into a Rumanian patrol and had a nice sharp fire fight which left us with a few prisoners and no losses. It also told the Rumanians that something unpleasant was afoot, and they spent the whole morning in combing the area with rifle and artillery fire. In this part of the world there was ample cover and we suffered no losses. We did not waste ammunition and reply to their fire, but we used our time in extending our reconnaissance of the enemy position and in drawing up, on the ground, complete plans for fire support during the attack. A mountain battery went into position to our left rear and numerous OPs were established and put to good use. The hours passed by and it was noon before Lieb struck. At the first sounds we advanced with the other units of the Gössler detachment.

Before the 2d Company advanced Lieutenant Grau swept the hostile position with heavy machine-gun fire from his slightly elevated positions. Our men broke from cover and charged down the hill with the fury of a swollen torrent. The Rumanian decided not to wait; our charge swept him out of his trenches in the saddle, and in a few minutes we had reached the Lesului. Our prisoner bag was small, for the Rumanian showed marked ability in being able to slip from our clutches and disappear up one of the many ravines that dotted the saddle. However, our seizure of the Lesului summit was not long delayed and we bivouacked there for the night. The 2d Company was pleased with itself for we had only one man wounded in this frontal attack.

After dark reconnaissance detachments were sent south to locate the enemy and to search for food. Up to this time we had been living on extremely short rations. The detachments returned during the morning of the twelfth and reported that they were unable to make contact with the enemy. They brought in a variety of cattle and we got our cook fires going in a minimum of time. Food and the bright November sun made us forget the cold night in the tent.

Sketch 13: The situation at Lesului.
View from the north.

Observations: The assembly area for the attack on November 11 was on a reverse slope some two hundred yards from the enemy position. The enemy erred in not using combat outposts to prevent us from approaching to within so short a distance of his main battle position. The troops remained in this assembly area for several hours and were subject to harassing fire most of the time. The attack itself was supported by machine-gun fire delivered at two hundred yards' range. The terrain was such that this was the only way of affording fire support.

The individual heavy machine guns first forced the enemy to take cover at those points where the assault platoons expected to break in. They kept up their fire while the troops covered the gap between positions and then they lifted and placed their fire on the rear portions of the hostile position. After a successful breakthrough, they followed quickly and supported the attack from improved positions in the elongated saddle. The enemy had been expecting an attack for several hours, but our mode of fighting was a total surprise to him.

We would have achieved an even greater success had we delayed the attack for thirty minutes. At that time Lieb's outfit would have been in their rear instead of being merely on their flank.

III: Battle at Kurpenul-Valarii

In the afternoon of November 12, 1916, the 2d Company, with a heavy machine-gun platoon attached, was ordered to move down the east slope of the Lesului and take the village of Valarii. At the same time the remainder of the battalion was to move down the western slope in two columns and attack the same objective. There was plenty of sunshine on the Lesului, but on our way down we ran into a heavy fog; I was obliged to feel my way down a valley-bound path by means of a compass. It was not long before we began to hear voices coming from the valley; we did not know if they were commands or merely conversation.

Not far below and to the left a Rumanian battery was firing at the Vulcan Pass. Our position was such that we might bump into the enemy in the fog at any moment. With strong advance, flank and rear guards, we felt our way down the grassy slope. All talking was forbidden.

By the time the fog lifted it was getting dark. Some thousand yards ahead in the valley we could see a long, narrow village consisting of single houses. Valarii or Kurpenul? Field glasses allowed us to distinguish small groups at various places, probably soldiers. Sentries were apparently stationed at the entrances to the village which lay within ten minutes' march of our halting place.

I considered it inadvisable to continue the march or to attack before establishing contact on both flanks or without awaiting the arrival of supporting units. My decision was to prepare to attack the village while awaiting the establishment of more complete flank liaison. To avoid giving our position to the enemy, I withheld all reconnaissance forward and decided to rely on keen visual observation.

I kept my outfit ready for an attack on the village in case our support came up before dark. We remained concealed in small depressions and clumps of bushes until dark when I ordered the organization of a hedgehog defensive position, sent out my security elements and sat down to await developments. All sentries were instructed to alert us as soon as the other units came up or whenever they heard any suspicious sounds. In this manner we got a few hours' sleep by lying on our arms.

Shortly before midnight we heard the flank units of our battalion coming down the slope. I alerted my men and, in the bright morning sunlight, we slipped through the underbrush toward the village of Kurpenul-Valarii with the heavy machine-gun platoon disposed on the left to give us fire support. The forward elements reached the edge of the village without difficulty, and reported that they could find neither hide nor hair of the enemy. On the other hand, shots occasionally rang out on the right near the neighboring column. We moved cautiously into the village and then brought up the machine-gun platoon.

The various farmhouses were inhabited with all members of a family sleeping around their fireplaces covered with blankets and furs. The air in those rooms was thick enough to be cut with a knife. We had considerable trouble in making ourselves understood by the natives. There were no signs of the enemy. A brief reconnaissance showed that we could convert the schoolhouse and the two adjacent farm buildings into a good strongpoint. We got to work, and after sending out the necessary security elements I took two runners and went to the western part of the village to find Major Sprösser and report. Other battalion units were settling down in the western part of the village from which the enemy had fled after the first exchange of rifle shots.

Major Sprösser divided the place into company sectors and we drew the eastern part of the village. We faced south with the 3d Company on our right. Liaison to the left was to be established with the 156th Infantry after daybreak. We were still ignorant of the enemy location and dispositions.

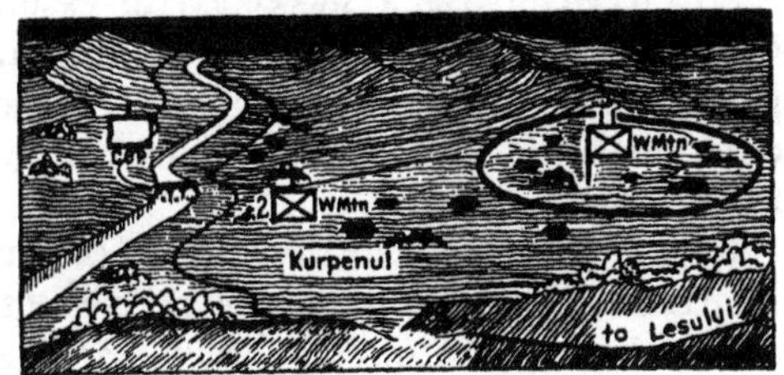

Sketch 14: The situation at Kurpenul. View from the north.

I returned to my company around 0300. It was a coal-black night. My men were sleeping in the schoolhouse. I awoke the subordinate commanders and we made a reconnaissance of our sector. Just east of our area a wooden bridge ran across the shallow Kurpenul, a brook some 150 feet wide whose banks were lined with poplars and weeping willows. There were roads leading south on both sides of the stream. The map showed the eastern one to be the better of the two. There were some farmhouses near the bridge and the village extended some hundred yards west of the brook. A dense fog, similar to the ones we had encountered during the preceding days, enveloped us before we had posted our security details which consisted of an NCO just west of the bridge and on the road leading through the village and combat outposts to the east of the Kurpenul. I also dispatched liaison detachments to establish contact with the 3d Company on our right and with the 156th Infantry on our left. When it finally grew light we found visibility restricted to sixty yards.

Before contact with our neighbors was established, Lance Corporal Brückner reported that he had run into a company of Rumanians about half a mile southeast of our combat outpost. The Rumanians were closed up and had bayonets fixed, but had not discovered Brückner's squad. Scarcely had I telephoned this report to the battalion than I received another report from the outpost at the bridge: "A Rumanian scout squad of six or eight men is in the fog some fifty yards to the rear of the outpost. Shall we open fire?"

While the company prepared for action, I hurried to the combat outpost. The high Rumanian fur cap made it easy to ascertain that we had enemy units roving in the area back of our outposts and I ordered a few company sharpshooters to open fire. We saw several drop with the first volley and the rest disappeared into the murk. A few minutes later lively rifle fire broke out to our left rear.

Other scout squads from the south reported that a strong Rumanian detachment was marching on the combat outpost east of the brook with its column head only a few hundred yards away. I quickly advanced one of my heavy machine guns toward the combat outpost and ordered it to sweep both sides of the road. This drew a few shots from the enemy and then all became quiet again.

So far we had not succeeded in establishing contact with the 3d Company (to the right), and to all appearances a several-hundred-yard gap yawned between the companies. We could hear lively shouting to our right which indicated that the enemy was advancing on a broad front toward Valarii-Kurpenul.

In order to close the broad gap between ourselves and the 3d Company, I started the company south along the west bank of the Kurpenul leaving the combat outpost and one heavy machine gun on the east flank at the bridge to protect our flank and rear. I wanted to reach the south edge of Kurpenul where I hoped to find a favorable field of fire, and to use the open ground to establish contact rapidly with my neighbor on the right.

I went with the advance guard, consisting of one squad, and the remainder of the company followed 160 yards to the rear. The fog swirled hither and yon and the visibility varied between a hundred and three hundred feet. Shortly before the head of the column reached the south end of the village, it ran into a close column of advancing Rumanians. In a few seconds we were engaged in a violent fire fight at fifty yards range. Our opening volley was delivered from a standing position and then we hit the dirt and looked for cover from the heavy enemy fire. The Rumanians outnumbered us at least ten to one. Rapid fire pinned them down, but a new enemy loomed on both flanks. He was creeping up behind bushes and hedges and firing as he approached. The advance guard was getting into a dangerous situation. It was holding a farmhouse to the right of the road, while the remainder of the company appeared to have taken cover in the farms some five hundred feet to the rear. The fog prevented it from supporting the advance guard. Should the company move forward, or should the advance guard retire? Since it was a question of asserting ourselves against a powerful superiority, the latter appeared to be the best thing to do especially in view of the extremely limited visibility.

I ordered the advance guard to hold the farmhouse for an additional five minutes, and then to retire on the right side of the road through the farms and reach the company which would furnish fire support from its position a hundred yards to the rear. I ran back down the road to the company; dense fog soon concealed me from aimed fire by the Rumanians. I quickly ordered a platoon of the company and a heavy machine gun to open fire on the area to the left and the advance guard began to drop back under this fire protection. The men were compelled to leave Private Kentner, who had been severely wounded, behind.

Figures loomed in the brook on our left and the stream was soon teeming with Rumanians. At the same time the combat outpost on the left became engaged in violent combat; its left flank was open and could be easily turned. On the right, at a considerable distance, another violent fire fight was in progress. We had not established contact with the 3d Company. If the enemy attacked on the right, the company would be completely surrounded. The tales the Bavarian soldiers told concerning the ascent of Hill 1794 (5920) came to mind. It must have been the same way with them!

My orders were: "1st Platoon holds the position under all conditions, 2d Platoon remains under my control behind the right flank of the 1st Platoon!" With a few runners I rushed off to the right to establish personal contact with the 3d Company. For some two hundred yards we ran behind hedges and across open fields. Just as we were crossing a freshly plowed plot of land we were fired on from a knoll some fifty to ninety yards on our right. The sharp reports were those of a carbine and that meant they were German. The furrows gave scant shelter and no amount of shouting and arm waving could convince them of the errors of their way. Luckily, their marksmanship was very poor. After a few anxious moments, a dense fog shrouded us and released us from this unhappy situation and allowed us to hurry back to the company. I gave up making further attempts at establishing contact with the 3d Company; I now knew where some of its elements were located and I hoped to be able to close the 280 yard gap with my reserve platoon. But as is frequently the case in war, things turned out differently.

On my return to the village street I discovered that, contrary to orders, the 1st Platoon and the heavy machine gun had attacked the enemy. Judging from the sound of battle, they had fought their way to the southern edge of the town. However praiseworthy the initiative of the platoon commander and his men might be, a defense of the southern edge of Kurpenul in the fog and against a superior enemy seemed hopeless without first establishing contact with the right or left. It was a good thing the reserve platoon remained in its assigned area.

The noise of battle increased and, suspecting the worst, I hastened forward to the 1st Platoon. Halfway there I met the platoon commander, who breathlessly reported: "1st Platoon has driven the Rumanians back three hundred yards south of the village and shot up two Rumanian guns. At the moment the platoon is very hard pressed by a strong enemy who is but a few yards away. The platoon is nearly encircled, the heavy machine gun is shot up, the crew dead or wounded. Help must come immediately, or the platoon is finished."

I was none too elated with this course of events. Why did the platoon fail to stay in its place as ordered? Should I commit my last reserves, as requested by the platoon leader? Under these conditions all of us might have been surrounded and crushed by superior numbers. Would such a loss have crippled the left flank of the Württemberg Mountain Battalion? No, as little as I liked it, I could not help the 1st Platoon.

I ordered the 1st Platoon to disengage immediately and fall back along the village road. The remainder of the company was disposed to cover the retirement of the platoon. Conditions for the disengagement from combat became more difficult for the sun was burning its way through the fog and visibility had increased to a hundred yards. Those were exciting moments. The 2d Platoon went into position in the middle of the village on the double and fired on the dense masses of the Rumanians who were attacking from the left front.

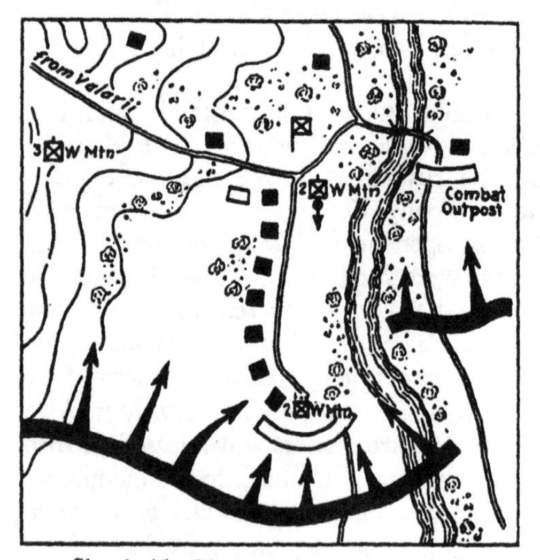

Sketch 15: The situation at Kurpenul.

Soon the remnants of the 1st Platoon began to shoot their way back, followed by a black mass of Rumanians. Rapid fire from the entire line brought part of the onrushing foe to a standstill, but on the right and left the hostile mass swirled closer. We now felt the lack of the heavy machine gun which had been lost up forward. The remnants of the 1st Platoon were rushed into the firing line. I hastened to the combat outpost on the other side of the bridge, found everything in order, took their available heavy machine gun and emplaced it in the most endangered place in the village.

But the Rumanians did not give way. In spite of great losses they attacked repeatedly. Company headquarters was put in the firing line. Its leader, Sergeant Dallinger, went down with a bullet in the head. The fog continued to dissipate and we had our first opportunity to judge the enemy strength. Our next worry was ammunition supply. The left flank was wide open.

I reported the situation by telephone to Major Sprösser and urged the immediate dispatch of additional forces. A few minutes later Lieutenant Hohl arrived on the double with about fifty men. I moved this platoon behind the left flank with the mission of defending the left flank with a few squads, leaving the bulk under my control. Soon afterward the 6th Company came up and was held at my disposal, echeloned to the left rear. There was nothing more to fear.

Meanwhile the 2d Company had dug in under fire. The enemy withdrew slowly under our well-aimed carbine and heavy machine-gun fire. I probed his front with scouts. We had good visibility and reached the southern edge of the village again where we found the severely wounded men of the 1st Platoon. They had been robbed by the enemy of their few belongings, such as pocket watches and knives, but otherwise nothing had happened to them.

The improved visibility showed the southern edge of the villages to be a splendid commanding position. I moved the company there, reorganized it, and began to dig in. Another heavy machine-gun platoon arrived.

The enemy had disappeared but we continued to receive rifle fire from the far left. On the right lay the battery which the 1st Platoon had destroyed. As it turned out later, other battalion units had also fired on it.

Since no enemy was in the forward area, I went up with a small patrol and took a look at the battery. Krupp guns! German workmanship!

Soon Rumanian skirmish lines reappeared in the south and approached our position. They were still over two thousand yards away, I gave the signal to fire at will. This stopped the attack cold and we suffered no losses in the ensuing fire fight. The heavy machine guns had many excellent targets. As night fell the enemy retreated. Company patrols took a few dozen captives in the forward area and the company prepared for the night. Advance scout squads failed to locate the enemy. The company dug trenches, and some soldiers looked around for a fat roast.

We were sad about the losses in the company which totaled seventeen wounded and three dead.

Like the 2d Company, the other units of the Württemberg Mountain Battalion stood fast at Valarii-Kurpenul, on the right wing of the Kühne group. They had had much to do with the complete success of this thrust across the mountain. On the Rumanian side hundreds of dead covered the field including a Rumanian divisional commander. This battle opened the road into Wallachia and we pounded on the heels of the beaten enemy. Two days later the Württemberg Mountain Battalion entered Targiu Jiu.

Observations; In a dense fog the reinforced 2d Company, on the afternoon of November 12, made the descent with security elements of all sides (advance guards, flank guards, and rear guard). The situation was most obscure, and the enemy might have been encountered at any moment. To spare the troops, an evening rest in combat formation (hedgehog defense, rifle at hand, scouts out front) was permitted.

The importance of combat reconnaissance and establishment of contact with the neighbors is impressively demonstrated by the events on November 13. Without the prompt knowledge of the advance of strong Rumanian forces, the reinforced 2d Company would have been crushed by the hostile mass in the fog.

The first combat outpost opened up with machine-gun fire in the direction of the advancing enemy. This clarified the situation rapidly and gave the 2d Company time to close the large gap on the right.

In the encounter of the advance guard with the hostile forces in the dense fog on the southern edge of Kurpenul, a bayonet fight did not develop, but a fire fight did. Why? With our inferiority in numbers a bayonet fight would have been inadvisable. We would have been cut and shot to pieces by the superior manpower of the enemy. But the rapid fire of a few riflemen hindered the attack of the tenfold superior enemy.

Both the advance guard and later the 1st Platoon shot their way through the fog back to units in position. In this they were very strongly supported by the fire these units delivered into the fog in the area between the village street and the Kurpenul brook which swept along close beside the line of retreat.

It is very easy to be fired on by one's own troops in combat in a fog. Here, as once before on the Brière farm, neither shouts nor signals managed to stop the fire.

The extremely difficult situation in the village battle against a very superior enemy was overcome by committing the last men at the focal point of the defense and by moving up forces from other less endangered places. The leader must be very active in such situations.

IV: Hill 1001, Magura Odobesti

In the middle of December we marched through Mirzil, Merei, Gura Niscopului, Sapoca, into the Slanicul valley, where we joined the Alpine Corps.

In the plains, the Rumanian resistance stiffened considerably thanks to Russian divisions which had been rushed in as reinforcements. The German Ninth Army slowly fought its way through Buzau to Rimnicul Savat and Fort Focsany. Our gains were at the cost of many casualties. The Alpine Corps received the mission to clear the enemy out of the almost impassable mountain area between the Slanicul and Putna valleys. This would relieve the forces fighting in the plain, and also prevent any hostile advance from the mountains against the forces operating against Focsany.

We spent Christmas Eve deep in the mountains under the most uncomfortable conditions imaginable. Then the 2d Company marched, in Alpine Corps reserve, from Bisoca through Dumitresti, De Long, Petreanu, to Mera. On January 4, 1917, we rejoined the battalion, whose staff was located in Sindilari. During the same afternoon, the company, reinforced by a heavy machine-gun platoon under the command of Lieutenant Krenzer, occupied Hill 627 (2070) about a mile and two thirds northwest of Sindilari. To cover Focsany, strong Rumanian formations held the extensive, rough and heavily wooded mountain range of Magura Odobesti (3300 feet altitude).

This mountain was to be taken on January 5. The Bavarian Infantry Life Guards were to be committed from south and southwest and the Württemberg Mountain Battalion from southwest and west.

My reinforced company had the mission of seizing Hill 1001 (3300) by attacking (without contact on either flank) across Height 523 (1725) (a mile and a half northeast of Sindilari). On the right we had the Bavarian Infantry Life Guards with their left wing about four miles to the southeast in the region of Hill 479 (1581). Lieb's detachment was on our left on the ridge leading to Hill 1001 (3300) from the west. He was some three miles from Hill 627 (2070). All these units had the same objective.

In accordance with orders, we moved forward at daybreak and, after crossing several deep and wooded valleys, reached Hill 523 at sunrise. An abandoned telescope rendered yeoman service. While the company rested under cover, I studied all the mountain slopes and valleys with the glass and soon became familiar with the disposition and strength of the opposing enemy forces.

Unfortunately the field of vision did not extend sufficiently to the right to locate the Bavarians on our right. In front of us (in a northeasterly direction) and about a thousand yards away, Rumanian reconnaissance detachments were patrolling the valley. Second, the ridge running in a north-south direction in front of Hill 1001 (3300) was completely occupied by Rumanians and sections of entrenched positions were clearly recognizable through gaps between trees. A covered avenue of approach through the broad, treeless valley in front of them was impossible by day. Over on the left, Rumanian combat outposts in about platoon strength stood on the ridge north of Hill 523, which was crowned by single farmsteads and small sections of woods. These outposts were located in entrenched positions facing generally to the west. The most promising avenue of approach to the Magura Odobesti was the ridge running from the west toward the summit along which Lieb's detachment was to be committed. I decided to move closer to Lieb's detachment and to operate in conjunction with him, since an advance in a northeasterly direction without contact to right or left against strong hostile forces seemed hopeless. To be sure, we were still three miles as the crow flies from Lieb, whom I was unable to see and whose presence was only to be presumed.

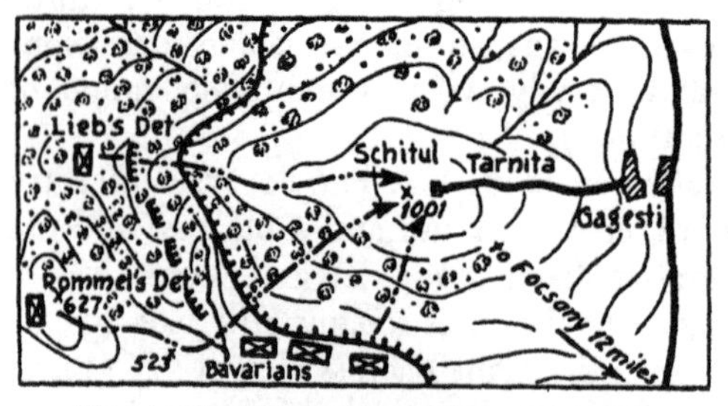

Sketch 16: The attack against Hill 1001. (Magura Odobesti).

I sent several reconnaissance detachments out with the mission of diverting the enemy's attention from my intended direction of attack (north) and instructed them to rejoin the company within two hours. Shortly after that we succeeded, without losses, in attacking the hostile combat outposts in succession and drove them back to their main position.

We reached a strip of wooded terrain and went to within a mile and a third of the ridge on which we supposed Lieb's detachment to be located. I turned off toward the north with the intention of gaining the ridge running in a north-south direction in front of the Magura Odobesti at the spot where it joined the ridge running from the west toward Hill 1001 (3300).

I marched out in front of the column with the company following 150 yards behind. In single file we passed through the sparse woods until we reached a cart road which went down into a ravine. When the scouts had reached the deepest part of the ravine, we noticed movement on the opposite steep slope. A Rumanian column with numerous pack animals was descending by zigzagging, with its head only a hundred yards away. Its strength was not discernible. What were we to do?

Apparently the enemy had not noticed us. Quickly I moved the point to one side into the bushes, then withdrew some fifty yards and placed my men in ambush. While this was going on, I sent a runner back to the leading platoon with the order to deploy. Before this was executed, Rumanian rifle fire began to strike among us. The point replied and in a few minutes the 1st Platoon joined the fire fight. Our position in the ravine was unfavorable, for the enemy, whose strength was hard to estimate, was firing from a superior elevation. In a long fire fight, heavy losses on our side were unavoidable. Therefore I decided it was best to attack the unknown forces. The result exceeded expectations. The enemy surrendered when our charge hit him and our bag included seven Rumanians and some pack animals. We suffered no losses.

We rushed up the slope after the retreating enemy and reached the crest out of breath only to be struck by heavy fire. On the left my brave runner Eppler fell with a shot in the head. After disposing the heavy machine-gun platoon and two infantry platoons, I attacked down both sides of the road in a northerly direction through the high forest. We advanced slowly, unable to see the enemy, the only evidence of his presence being the strong fire humming about our ears. To all appearances the fire became stronger the farther we advanced. Finally we found ourselves lying in a sparse high forest some three hundred yards from a fortified position. Resistance was so strong that further attack seemed hopeless. A shallow saddle separated us from the hostile position and our position on a forward slope was unfavorable.

To avoid unnecessary losses, I ordered the riflemen to withdraw to the next hill under cover of the heavy machine-gun platoon. This maneuver was executed and we found ourselves about a quarter of a mile from the enemy who was occupying a small knoll. The firing died out by degrees and soon only occasional shots were to be heard.

Having no contact on either flank, we formed a hedgehog and began to dig in with the reserve and heavy machine-gun platoons in the center of our defense area.

Before complete darkness set in we had located elements from Lieb's detachment over on our left on the edge of a glade some eight hundred yards away and had established wire communication with them.

I discussed the situation with First Lieutenant Lieb and later with Major Sprösser. A frontal attack by the two detachments against the strongly fortified Rumanian forest position offered small chance of success. The possibility of an envelopment from the southeast had to be determined without delay.

During the night Technical Sergeant Schropp made a thorough reconnaissance of the south flank of the hostile position—an extraordinarily difficult task in the rugged terrain. A few hours before daybreak he brought back this excellent bit of information: "We moved off to the northeast, crossed a deep ravine, and managed to reach the ridge behind the enemy position without encountering any hostile forces. Then we crossed a road which is apparently carrying heavy Rumanian traffic."

I reported these results to Major Sprösser and was ordered to execute the envelopment with two and a half companies. Daybreak was given as the time of attack. Lieb's outfit was to execute a frontal attack only after my unit had launched its attack. At this moment it began to snow heavily.

A gloomy day broke on a four-inch blanket of snow. Snow clouds covered the heights. The 6th Company came up as reinforcement. I left Hügel's infantry platoon behind in the old position with the mission of pinning the enemy down by frontal fire and of distracting his attention from us. With one and two-thirds companies and the heavy machine-gun platoon I moved off to the east and climbed down into a very deep ravine. Schropp led since he had been over the route during the night.

Hügel opened fire from our old position and had a lively response from the Rumanians who apparently feared an attack. While the fire fight was on, we silently crossed the ravine and climbed in a northeasterly direction. After a hard ascent we gained the ridge and came upon a fresh path in the snow made by Rumanian detachments.

Fog had reduced visibility to less than fifty yards and we expected to run into the enemy at any moment. I ordered the 2d Company to drop packs and rapidly organized the detachment for attack. The 2d Company and the heavy machine-gun platoon led with the 6th Company in the second line at my disposal. Except for occasional shots Hügel's fire on the left had died out.

We moved carefully forward astride the ridge road and through the wintry forest toward the enemy west and rear. Suddenly we heard voices before us in the fog. I halted and had the heavy machine gun prepare to open fire. Then we slipped cautiously forward. The camp fires were still smoking, but no Rumanians were to be seen. We went on until we came upon a clearing in the woods where we saw several unsuspecting Rumanians moving about. How strong was the enemy? We did not know whether we were opposed by a few individuals or by a whole battalion. Preparing for any eventuality, I ordered the heavy machine-gun platoon to open fire on the figures moving in the fog. A few seconds later my whole detachment rushed toward the enemy with loud shouts.

Only a few Rumanians were there and they elected to seek safety in flight rather than to stand and fight. We did not bother with them, but raced along the road to the west. We began to receive fire without being able to locate the enemy, and then after a few minutes we heard the approaching shouts of Lieb's detachment.

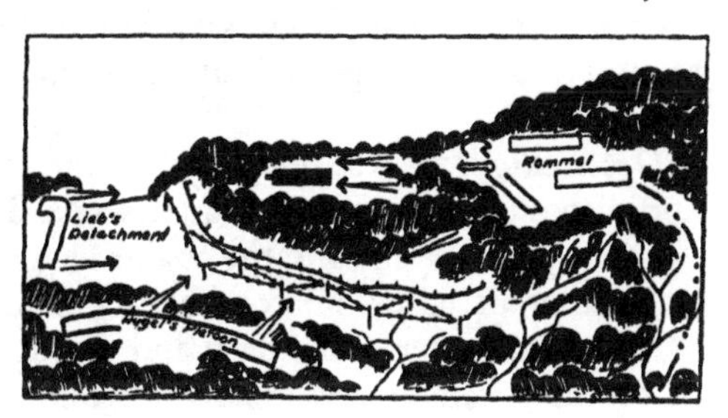

Sketch 17: Situation of January 6, 1917.
View from the south.

We had to be careful to avoid firing on Lieb's men approaching in the fog and forest. We solved this ticklish problem and the enemy between our detachments was eliminated. Most of the Rumanians avoided immediate capture by fleeing downhill and the 2d Company only gathered in a total of twenty-six prisoners. They only postponed their fate. Three days later when our outfit was already on the Putna an entire battalion of five hundred men emerged from the woods and surrendered in a body to the commander of a pack train.

Following our successful attack, which involved no losses on our side, Lieb's detachment headed toward Hill 1001 (3300). I ordered the 2d Company to pick up its discarded packs and then joined the advance. Snow began to drift and the fog became denser.

Near the summit of Hill 1001 (3300) Lieb met some Rumanian reserves who had taken up a position in a place sheltered from the wind. The resolute attack of our mountain troops made quick work of them and the Rumanians abandoned the hilltop after suffering some losses. They did not return to their snow-drifted positions.

A cold wind swept over Hill 1001 (3300). Ice crystals stung our faces like needles. These weather conditions forced us to hurry our units to the shelter of Schitul Tarnita monastery which was located a short distance down the east slope of the mountain. The enemy did not block our way. To be sure, the monastery failed to meet our expectations especially as space and rations were concerned, but it at least offered protection against the inclemencies of the weather. Unfortunately our joy was short-lived.

An hour later some units of the Bavarian Infantry Life Guard arrived at Schitul Tarnita and claimed the monastery as their quarters. The Bavarians were our seniors and we had to give in. The Bavarian officer outranked Lieb and myself and we were obliged to move. Lieb managed to keep his people on the monastery proper, but my men had to find shelter in the woody, low-roofed, unheatable earth huts that lay near the monastery. We spent a miserable and bitterly cold night and I decided to push on as soon as possible and find the inhabited region of the valley.

Observations: It was possible to locate and study the hostile positions and dispositions by means of a telescope. This was done during the advance of the company and the results obtained were of equal importance to those prepared by our combat reconnaissance detachments.

In the encounter in the deep and wooded ravine, the vigorous assault of the mountain troops more than compensated for the poor tactical position.

By evening our attack had been stalled some three hundred yards short of the fortified battle area of the Rumanians. To avoid losses, I ordered the rifle platoons on the forward slope of the sparse forest to shoot their way back to a more favorable position under the fire protection of our heavy guns. No losses ensued. In a similiar situation efficient use could be made of smoke screens. Initially the enemy maintained a heavy fire into the smoke, but his inability to achieve definite results obliged him to suspend firing. This was the moment to begin disengaging operations.

The excellent results of the winter night combat reconnaissance (Technical Sergeant Schropp) made the advance to the rear of the enemy on January 6, 1917 possible. Principle: *Reconnaissance must be active while the troops are resting.*

To deceive, divert, and pin the enemy down during our envelopment, it was necessary for Hügel to carry out his fire mission for an extended period of time.

During the final phases of the envelopment when we launched an attack in the fog against an enemy of unknown strength, we placed our heavy machine guns well forward and their fire soon cleared the enemy from the ridge.

While the wind was piling the snow in drifts, the Rumanian reserves remained in a protected part of the slope of Hill 1001 (3300). This location was such that they were without communication forward and they had neglected to post security elements. Because of this Lieb's detachment had little difficulty in surprising and dispersing this strong enemy force.

V: Gagesti

Very early on January 7, 1917 I sent scout squads toward the Putna valley on both sides of Gagesti. It was bitterly cold with twelve inches of snow on the ground and a heavy fog prevailed. Toward 1000 Mess Sergeant Pfäffle reported that he had ridden about two and a half miles in the direction of the valley without encountering the enemy. At that point he heard sounds of numerous columns and much noise coming from the valley. Apparently the enemy was withdrawing although the fog prevented visual observation.

I forwarded this report to Major Sprösser by telephone and asked permission to take the 2d Company (reinforced) and feel out the way to Gagesti.

An hour later we moved off down the valley in single file through the sparse forest. Fog limited the visibility to about a hundred yards. Our security detachments consisted of advance and flank guards, the former made up of a squad under our able Technical Sergeant Hügel preceding us by about a hundred yards. The heavy machine-gun platoon was in the center of the company with its guns loaded on pack animals.

It took us thirty minutes to emerge from the woods and we found ourselves on a narrow footpath leading through a very dense nursery of young trees a few yards tall. I marched at the head of the main body. The fog had lightened.

Suddenly shots rang out in front followed by Hügel giving orders and then his report that we had encountered a Rumanian scout squad on the trail. His first shots killed the leading Rumanians and the remainder, seven in number, surrendered. Meanwhile the company had deployed for caution seemed indicated. Perhaps the prisoners were the security elements of a hostile column. Hügel continued his advance and, in a few minutes, reported that he had reached the eastern edge of the nursery and that an approaching hostile skirmish line estimated as a company was about a hundred yards away. Immediately I ordered the leading platoon to deploy at the nursery's edge astride the trail and open fire. Our answer was violent hostile fire which whistled through the brush and drove us to earth. The employment of the heavy machine-gun platoon caused some difficulty and its leader reported that his guns were frozen and that he would have to thaw them out. A lively exchange broke out a few yards to the east of the nursery edge. To all appearances we had met an enemy force in superior strength. In a small hollow the heavy machine-gun platoon was working feverishly to thaw its guns with alcohol. The hostile fire rattled through the low trees. It was more than annoying that the heavy machine guns were unable to intervene at that moment. If the enemy had enveloped our left or right, we would have been obliged to retreat. The 2d and 3d Platoons provided security in these directions.

Finally the first machine gun was in order and went into position, but it never had an opportunity to fire.

In an ever-thickening fog the enemy disengaged from combat and soon deprived us of remunerative targets. To fire into the fog would have been a waste of ammunition, improper for mountain troops operating under our difficult supply conditions. Under the fire protection of the heavy machine guns I took a platoon and advanced to a slight elevation crowned by a small house standing in a fenced-in vineyard. No shots were being exchanged. We could see many Rumanians milling about in a leaderless fashion on the bare slope across from us. We waved to them with our handkerchiefs and soon had twenty prisoners without firing a single shot. The Rumanians were sick of a war which had certainly gone very badly for them. Some of the prisoners helped us to round up more of their comrades. The rest of my company came up. Our position was such that the enemy was capable of striking us from any point of the compass. Therefore we prepared the position for all-around defense with security and scouting elements out in all directions. These began to send more prisoners back. Lance Corporal Brückner surprised five Rumanians in a vineyard building and quickly disarmed them. Lieutenant Hausser and I went into the forward area in search of a more suitable spot in which to locate the company. We hoped to find a farmstead. The temperature was 15 degrees and we were beginning to suffer from the cold and hunger. We were unable to locate a farmstead in the vicinity, but we did find a better position for the company just north of a deep gully in the middle of a fenced-in vineyard. A small house was located in the middle of the position and here in a single, unheated room we found a severely-wounded Rumanian who had been abandoned by his countrymen. Dr. Lenz did what he could for him, but there was little chance of bringing him through. The company moved in.

The deep gully led down the valley toward Gagesti. The terrain to the north and east was open for about a hundred yards with a light brush stretching out in the other direction. Fog still swirled hither and yon and at times we had visibilities up to two hundred yards. We heard the sound of voices over on the slope to the left. Dr. Lenz and I crept in that direction and, some thousand yards from our own position, we discovered a large Rumanian troop formation, about a battalion in size, resting in an open field behind an orchard. Hundreds of men, horses, and vehicles were assembled in this small area. Campfires gleamed.

While the fog allowed us to approach without being seen, I decided against an attack because the terrain was such as to make it impossible to use our weapons with maximum effectiveness.

It was 1400 and and we had an hour and a half until dark. The extreme cold made it impossible to bivouac in the open. Where was Gagesti? We preferred to seize some village buildings for our night quarters instead of returning crestfallen to Schitul Tarnita. In addition to shelter we needed food. Hunger makes soldiers enterprising.

With Dr. Lenz and his orderly I moved to the east of the company position, on the left bank of a gully about ten feet deep. Technical Sergeant Pfeiffer with three or four men kept abreast of us some fifty yards on our right.

We had covered less than a quarter of a mile when we located a large number of Rumanians on the northern side of the gully near a small house. Were they a combat outpost? In spite of having but one carbine with us on the north of the gully and having but four on the south, we advanced toward the enemy and by shouting and waving handkerchiefs ordered him to surrender. The Rumanians neither budged nor fired. We were within thirty yards of them and retreat was out of the question. I was secretly worried over the outcome. The Rumanians were standing close together with their rifles at the order and were talking and gesticulating among themselves, but refrained from firing as if to show their friendly intentions. Finally we came up to them and had them disarm. I told them a cock-and-bull story about the end of the war and then turned the thirty prisoners over to Pfeiffer's squad.

The three of us continued eastward toward the valley. Some distance farther on we saw the outlines of a deployed company loom up out of the fog. They were still fifty yards away but we decided to risk it. We advanced waving our handkerchiefs and shouting. The company was taken aback. Their officers shouted angrily: "Foc! foc!" (presumably Rumanian for "Fire!") and also began beating their men, who apparently preferred to lay down their arms. We were in a most precarious position. The company took aim and a hail of lead whistled past. We dropped to the ground and then Dr. Lenz and I rushed to the rear while the doctor's orderly fired a few shots before taking leave. The fog soon concealed us from further aimed fire. A portion of the enemy followed us, while others fired at random into the fog.

Hard pressed by the enemy we reached Pfeiffer's squad and found the thirty prisoners still standing alongside their weapons. We herded them quickly into the gully, which offered cover against the pursuers' fire and chased them toward our company at the double. We would have been forced to abandon our gully had the enemy shot up its axis. The Rumanians were poor marksmen and we reached our outfit with all our prisoners and without any losses.

Soon after our return, the fire of the company halted the enemy who was advanced on a broad front. A lively exchange developed at a hundred yards' range and thanks to our heavy machine guns we enjoyed a considerable superiority in fire power. I decided that even a successful attack under existing conditions would fail to compensate for probable losses. Night was falling, and the intensity of fire was dying down with both sides firing intermittently to show they were still there. In the bitter cold the prospects of finding accommodations for the night and a warm meal were far from promising. Lieutenant Hohl (3d Company) arrived on horseback to see about us, took charge of our eighty prisoners and moved them to the rear. He also reported in Schitul Tarnita that I had decided to make a night advance on Gagesti.

During the past hour the weather had cleared considerably but the cold had also intensified. The stars shone in the sky and the bushes and trees were black silhouettes against the white snow. Carbine and machine-gun fire constituted my final greeting to the enemy and then I disengaged my force. We moved silently up the narrow mountain path in a northwesterly direction. Advance guard and rear guard secured the march and the heavy machine-gun platoon was in the center of the column. The heavy machine guns, still warm from firing, were protected from freezing by blankets and shelter halves. After proceeding some six hundred yards on the path, I turned off to the north. The North Star replaced the compass and we slipped forward along black thorn hedges which allowed us to move without standing out in contrast to our surroundings. Not a word was spoken. The rear guard reported that a strong Rumanian detachment was following it, whereupon I halted at a dark row of bushes and set up a heavy machine gun. The maneuver proved to be superfluous, for the leader of the rear guard acting on his own initiative ambushed the enemy at a suitable spot, and captured him without a shot. Twenty-five Rumanians! They were of no use to me so I sent them under guard to Schitul Tarnita.

We moved on to the north. A half mile farther on I turned again to the east. Before moving out I had studied the map thoroughly. We had come out on a dead line with the north end of Gagesti. The company deployed silently and advanced with all three platoons abreast; I was with the heavy machine-gun platoon in the center. Thus we felt our way from bush to bush. The land sloped gently toward the Putna valley. We halted repeatedly and carefully observed the surroundings with the glasses.

While the moon rose on our right, the glow of a fire became visible on the left in the valley before us. Soon we located several dozen Rumanians standing about a huge campfire some seven hundred yards away. Beyond, a hostile detachment was marching from left to right, presumably toward Gagesti. The village was concealed by a long, bare hill, on which only individual clumps of trees were to be seen with the glasses. To the right front the view was cut off by extensive orchards.

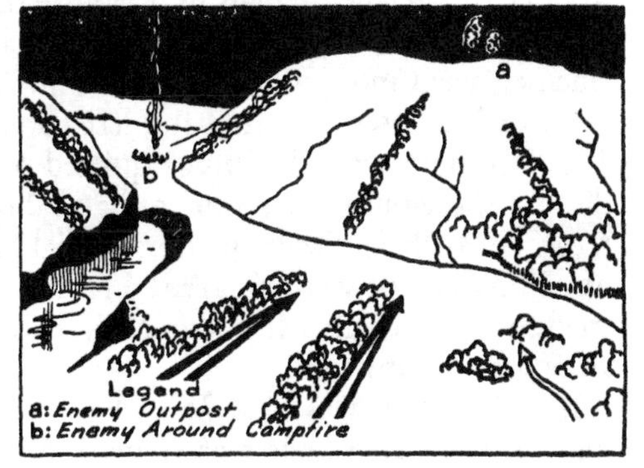

Sketch 18: In front of Gagesti.

Like hungry wolves the mountain troopers crept closer in the cold winter night. Was I first to attack the enemy to the left front in the valley, or was I to by-pass him and head straight for Gagesti?

The latter course seemed best. Clinging closely to the black hedgerows, the three columns crept slowly and cautiously forward until they were within three hundred yards of the bare hill whose summit was still a hundred feet above us. About fifty Rumanians were sitting around a fire three hundred yards on our left. Several of my men claimed to have seen signs of movement among the groups of trees on the hill ahead of us, but I was unable to verify these observations with my field glasses.

We crept along the hedges and finally gained the lower part of the hill, which could not be observed from above. During the time required for assembly, scouts worked their way forward to the crest of the hill where they located Rumanian sentries some hundred yards ahead of us. The first question was whether I should await the heavy machine guns. That seemed unnecessary for the few men involved. I wanted to seize the hill by surprise and, if possible, without resorting to gunfire. The attack on the northwest part of Gagesti, which I assumed to be heavily occupied, was also to come as a surprise.

The subordinate commanders received their instructions and we rushed forward without a sound. Not a whistle, not a command, not a shout! The mountain troops rose before the Rumanian sentries as if conjured up out of the ground. It happened so rapidly that the latter did not even have time to fire a warning shot. They hurriedly disappeared downhill.

The hilltop was ours. Ahead and to our right front the moonlight glistened from the roofs of Gagesti, a village about half a mile long.

The nearest farmsteads lay a bare two hundred yards away at an elevation differential of a hundred feet. There were large intervals between the groups of buildings.

Alarm bells began to ring in the north of Gagesti. Soldiers rushed out into the street and gathered in clusters. At any moment I expected them to storm up in a dense mass to recapture the lost height. We were ready for them. The heavy machine guns were loaded for steady fire and the riflemen went into position on a two-hundred-yard front. One platoon remained in reserve behind the left flank.

The minutes passed. Things quieted down in the village. Since we did not show ourselves on the hill and did not fire, the alarmed troops returned to their warm quarters, which they had probably left most unwillingly. We were amazed! For not even the Rumanian sentries tried to return to their former places. Apparently they were down there among the farmsteads.

By this time it was 2200. We were freezing and hungry and within sight of the warm houses of Gagesti. Something had to happen. The decision: The northern-most farms of the large village were to be captured from the enemy. We would entrench ourselves in them, warm and feed ourselves, and at least rest until daybreak.

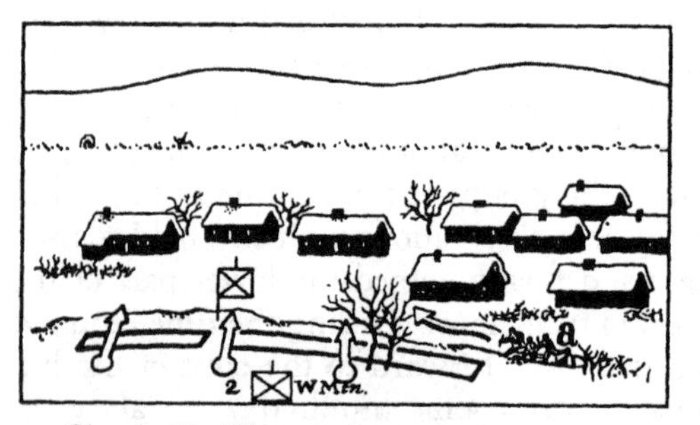

Sketch 19: The attack on the farmstead.
(a) Hügel's assault team.

I sent Technical Sergeant Hügel with an assault detachment of two squads from the right platoon against one of the farms. He was to proceed along a dark hedge; if fired on, he was to answer the fire and then in conjunction with the left platoon, capture the farm opposite him under the fire support of the other units of the reinforced company. The various units were instructed as to their tasks, and Hügel moved forward.

The assault group approached to within fifty yards of the farm before it was fired on. All the machine guns and Janner's platoon opened fire immediately and the left platoon rushed toward the village with a great shout. The mountain troops were in the town. Hügel attacked from the other side before the Rumanians could get out of the buildings. The rest of the reinforced company roared out into the night at the top of its lungs, as loud as a battalion. The heavy machine-gun platoon could no longer fire without endangering our own troops in the farms on the north end of Gagesti and shifted its fire to the right and sprayed the roofs of the extended village for several minutes.

Down below the north end it became surprisingly quiet. Only a few shots were being exchanged. The Rumanians were surrendering. I hurried in that direction with another platoon and a heavy machine-gun platoon. The prisoners were being gathered together as I arrived among the buildings. There were more than a hundred of them. Still more pleasant was the fact that none of us had been wounded during the fire fight. Not a shot came from the farms round about. Only our machine-gun platoon occasionally fired to the right across the roofs. Since everything had gone so well, I moved to the right with the company from farm to farm. We captured the entire Rumanian garrison which accepted its fate without offering further resistance. With all-around security and with the prisoners and the platoon in the center, I moved with the whole company toward the south along the village road. Two hundred prisoners! There was no end to them. Mountain troopers knocked at the doors all around and brought out new prisoners. We approached the church. The number of prisoners was three times our own. Three hundred and sixty men!

The church was on a small elevation, which descended sharply to the east toward the lower village some two hundred yards away. A semicircle of dwellings lay about the church. This seemed to me to be the ideal place for safe accommodations for the remainder of the night. The prisoners were herded into the church and the company was quartered in the houses round about. I made a reconnaissance of the lower village, through which the Odobesti-Vidra road passed, but encountered no more Rumanian soldiers. To all appearances the sound of combat in the upper village made them shift their quarters to the east bank of the Putna. I met the local mayor, who through a German-speaking Jew informed me that he wished to surrender the keys of the town hall. In anticipation of the arrival of the German troops, the community had baked three hundred loaves of bread, slaughtered several head of cattle, and placed a number of casks of wine at the troops' disposal. I had them bring enough for our needs to the church in the upper village which we had made our quarters. It was past midnight when the last units of the reinforced company moved in. Sentries protected the sleeping men.

Being located some four miles ahead of our own lines without contact to right or left, I felt safe in Gagesti only so long as it was dark. For safety's sake, I wanted to be on a commanding height just east of Gagesti by daybreak, at which time the enemy would be definitely located.

The troops ate and rested. I prepared a short report, which was dispatched at 0230 to Schitul Tarnita by runner. He also took along a *Logel* (a three-quart wooden measure) of exquisite red wine for First Lieutenant Lieb.

The rest of the night passed without disturbance. Just before daybreak (January 8) I moved my whole formation to the heights just east of the church at Gagesti. When day broke we were able to determine the snow-covered territory round about to be free of the enemy. We did see enemy troops entrenching on the east bank of the Putna. I returned to our former quarters around the church and sent out scout squads in various directions.

Mess Sergeant Pfäffle and I took a morning ride through the lower village in the direction of Odobesti. During the night we had sent our pack animals back toward Schitul Tarnita for their neighing would have betrayed our advance from Gagesti. Pfäffle brought the remainder of the detachment up after daybreak. I rode in the direction of Odobesti in an effort to establish contact on the right with our own troops west of the Putna.

Not a shot sounded as we trotted through the lower village of Gagesti. A ride in the cool of morning was most refreshing. I let "Sultan" step out briskly and paid more attention to the horse than to my surroundings. Pfäffle rode about ten yards behind me. We were about eleven hundred yards from Gagesti when something moved on the road ahead of my horse. I looked up and was more than surprised to see a Rumanian scout squad of about fifteen men with fixed bayonets right in front of us. It was too late to turn back and gallop away, for any indication of intended flight would have brought me a couple of bullets. I quickly made up my mind, trotted up to the scout squad without changing pace, greeted them in a friendly way, gave them to understand that they must disarm, that they were prisoners, and were to march toward the church in Gagesti, where four hundred of their comrades were gathered. I doubt very much whether any of the Rumanians understood my words. But my demeanor and my calm, friendly tone of voice had a convincing effect. The fifteen men left their weapons on the road and moved off across the fields in the indicated direction. I continued my ride for another hundred yards and then galloped back to my company by the shortest route. I would probably not have encountered such simple adversaries a second time.

In the course of the forenoon the 1st Company and 3d Machine-Gun Company arrived as reinforcements and were attached to my command. The Rommel detachment now consisted of two rifle companies and a machine-gun company. Lieutenant Hausser was adjutant.

Our scout squads brought in more prisoners. Toward 0900 "the war started again." Rumanian and perhaps Russian artillery subjected Gagesti to very lively harassing fire from positions on the heights east of the Putna. We vacated the most endangered places for we had plenty of room in the extensive village. Fortunately we suffered no losses.

During the afternoon the hostile fire increased to great violence reminding us of the western theater of the war; shells fell all around. Shots came through the roof of the house in which the detachment command post had been established. Here too—as often before—the violent bombardment was probably the result of the active movement of runners. Conditions became very uncomfortable. The detachment occupied the outskirts of Gagesti and dug in. Was the enemy going to attack?

During the heaviest phase of the bombardment Major Sprösser arrived in Gagesti on horseback and set up his command post in the front line along the Odobesti-Vidra road. The hostile artillery continued firing with undiminished violence until dark. We figured on a night attack, of which the Russians are especially fond, and secured our open flank with especial care.

Observations: A few shots quickly decided the battle in the forest nursery between the advance guard and the Rumanian scout squad. At such times it is important to move toward the enemy with weapons at the ready (safety off, light machine guns carried in position for shooting). For he wins who fires first and can deliver the heaviest fire.

In the fire fight a few minutes later with a stronger enemy, the heavy machine guns froze at the most critical moment. They had to be heated with an alcohol flame a few yards behind the front line. During later phases the heavy machine guns were kept warm with blankets.

The disengagement from fighting at dark was achieved without friction after a short, powerful burst of fire on the nearby enemy.

The night attack on the north part of Gagesti by moonlight over the snow was from two directions with strong fire support by the heavy machine-gun platoon. Even after the successful attack this platoon supported the advance in the long village by indirect fire delivered over the tops of the houses. There was little to be hit, of course, but the psychological effect on the enemy within the warm quarters was so great that he allowed himself to be captured without offering much resistance. There were no losses on our side in the fighting at Gagesti.

VI: At Vidra

At midnight we were relieved by units of the Alpine Corps and, in bright moonlight, moved to the north over the valley road. We marched seven miles, sometimes eleven hundred yards in front of the newly created Rumanian and Russian positions, without being attacked. Our troops were not opposing the enemy here. At daybreak the staff of the Württemberg Mountain Battalion and the Rommel detachment arrived at Vidra where we found comfortable quarters for the first time in days.

I was just making myself comfortable when the following battalion order reached me: "Enemy has broken through in the mountains north of Vidra. Rommel detachment prepares to move to Hill 625 (2070 ft.) north of Vidra, where it is attached to the 256th Reserve Infantry."

This demand was almost beyond human strength. For four days my detachment had been fighting under the most difficult conditions and had just completed a hard night march. The dead-tired soldiers had just moved into their quarters. They were to be thrown into battle on the snowy mountains north of Vidra.

At the assembly area I told the companies in a few words about their new task. Then the detachment moved north into the mountains. I galloped ahead with Lieutenant Hausser, Sergeant Pfäffle, and a mounted runner. The untiring legs of the horses quickly carried us over long, snowy mountain meadows and into the danger zone.

There were adequate reserves available and my detachment was not committed. We received the battalion order to return to Vidra after a cold night around campfires in the deep snow. In gay spirits the troops moved toward the comfortable billets where mail from home awaited us.

The Württemberg Mountain Battalion was at the disposal of General Headquarters and in the following night moved—again marching past the hostile front at Gagesti—back to Odobesti. In the following days we marched through Focsany fortress, which had fallen in our hands and Rimnicul Sarat reaching the vicinity of Buzau.

Rail transport was tied up because of severe snowstorms, but we finally entrained and headed west. We had a ten-day ride in unheated cars. It was bitterly cold. In the Vosges we went into army reserve for a few weeks, then we moved forward to the Stossweiher-Mönchberg-Reichackerkopf sector.

A third of the battalion (2 rifle companies, 1 machine-gun company) became corps reserve in Winzenheim and remained under my command. Major Sprösser instructed me to use this period to restore the old level of combat efficiency. This meant training and combat exercises. This task was most congenial. In the course of the next few weeks all companies of the battalion passed through my school. The curriculum was varied and designed to keep the troops ready for action. The program included night alarms, night marches, attacks on prepared positions, and all forms of combat that a German soldier might expect to face.

In May 1917 I took over a small sector of the Hilsen Ridge. In the beginning of June the French hammered us for two days on a wide front and the positions built up by more than a year's work were leveled inside of a few hours. But the hostile infantry attack failed to materialize. Our protective fires apparently stifled his attacking order. The battalion was recalled to new duties before improvements and repairs were completed on the demolished positions. Full of desire for achievement, the troops, probably then at the very peak of development, left the high Vosges. Once again the favorite song of the Württemberg Mountain troops, *Die Kaiser-Jäger,* resounded through Winzenheim.

Chapter 8

FIRST OPERATIONS AGAINST MOUNT COSNA

I: Approach March To The Carpathian Front

Although the outbreak of the Russian Revolution weakened the Allied position on the Eastern front, the summer of 1917 found large German forces still pinned down in that area. Nothing short of a complete eradication of the entire front would release these forces for the final decision in the West. To this end the southern flank of the Russo-Rumanian front was to be attacked from the south by the Ninth Army, which was located between the lower course of the Sereth and the edge of the mountains twenty miles northwest of Focsany, and from the West by the Gerck group, which was in contact to the left in the mountains.

After a week's train ride in intense summer heat from Colmar via Heilbronn, Nürnberg, Chemnitz, Breslau, Budapest, Arad, and Kronstadt, the troop train under my command (1st, 2d, and 3d Companies) arrived in Bereczk toward noon on August 7, 1917. We were the next to the last unit of the battalion to arrive. At the station I learned that the attack by the Gerck group was scheduled for the morning of August 8 against the heights on both sides of the Ojtoz valley. (See sketch 20.)

The three companies drew canned rations and, minus our trains, we took a three-hour truck ride over the Ojtoz pass to Sosmezö, which was near to the Hungarian-Rumanian border of that time. The combat and provision trains were to move up to Sosmezö as soon as they had unloaded.

In Sosmezö we met the valley detachments of the battalion which had marched to the mountains north of the Ojtoz valley during the forenoon. Telephone connection with battalion headquarters had been interrupted and a commissary sergeant transmitted the battalion orders orally: "Rommel's unit follows the battalion as soon as possible to Hill 764 (2620) (Bolchan) via Harja—Hill 1020 (3370)."

Austrians, Hungarians, and Bavarians occupied the valley in force and many batteries, some of major caliber, lined both sides of the valley road. Since I could not start the march into the mountains until the combat train arrived, I ordered the unit to bivouac in a very small area.

Austrian sentries with fixed bayonets watched that none of my riflemen got into the local commandant's potato patch. This precaution was justified because of the extreme ration shortage existing at that time.

Night fell and the battalion band gave an hour's concert among the camp fires. Our memories of the last winter's campaign in Rumania caused us to view the future with great confidence.

The fires were extinguished at 2200. The troops slept, which was necessary, for the ensuing days would certainly demand the greatest exertions.

The night's rest lasted a few short hours for the combat train arrived at midnight. Shortly thereafter I gave orders to wake up, break camp, distribute four days' provisions, and get the companies ready for the road. Since all vehicles remained in Sosmezö, the companies and the detachment staff took a few pack animals from their respective trains to carry ammunition, provisions, and baggage. Then the unit began to march via Harja. The column moved forward noiselessly in the clear, warm, moonlit night. By daybreak I wanted to clear those portions of the valley and Hill 1020 (3370) that were presumably under enemy observation. From Harja the steep and slippery way led mostly through forests. At daybreak the companies had a chance to prove their mettle by pulling an Austrian battery which was to take part in the battle up the hill.

In the course of the forenoon the artillery of both sides did a great deal of firing. We were afraid that we might be late for the breakthrough by the 15th Bavarian Reserve Infantry Brigade to which the Württemberg Mountain Infantry Battalion (WGB) had been attached. In spite of a very fast pace, it was noon before we arrived at the wooded Hill 764 (2520).

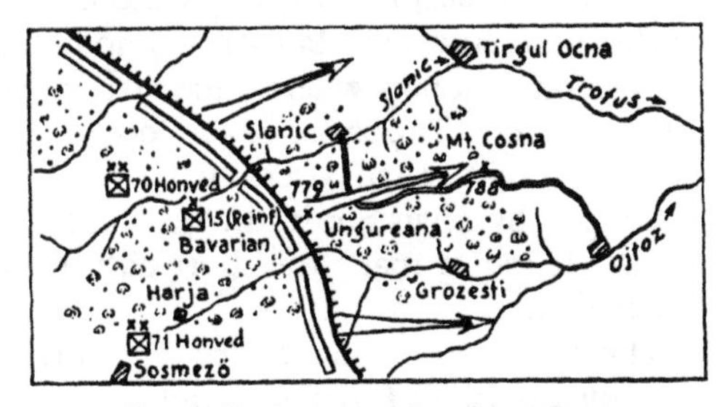

Sketch 20 : Attack against Ojtoz Pass.

While the unit was resting, I reported our arrival by telephone to Major Sprösser and received orders to move forward as brigade reserve to Hill 672 (2230) where Sprösser's headquarters was located. On arrival, I was given six companies and later three additional machine-gun companies. Concerning the course of the battle we learned that the 10th Bavarian Reserve Infantry Regiment had taken the first Rumanian positions on the Ungureana after a very stiff struggle. The Rumanians were said to have fought very bravely here, contrary to all expectations, and to have defended every trench and dugout with extreme tenacity. A breakthrough on the hostile front was not achieved.

My force was in position for the night, had pitched tents, and was cooking supper when orders came to move farther forward with three infantry companies and one machine-gun company to a point just west of the Ungureana (775) (2671). Major Sprösser went on ahead and

I followed with my four companies. It was pitch dark in the woods as we slogged along in single file over a swampy, narrow path. Flares were going up on the ridge ahead of us, machine guns chattered from time to time, and shells burst. We soon reached our destination. I reported our arrival and received orders to camp for the night in the hollows just north of the main trail.

The individual leaders had just been assigned their places and tasks, and the unit was still standing in a long line on the narrow path, when shells began to strike on the slope to the right and left. Surprise Rumanian concentration! On all sides the flashes of the bursting shells lit up the night, splinters whistled through the air and earth and stones rained down. Pack animals broke loose and stampeded into the dark with their loads. My infantrymen flat on the slope bore the fire patiently until the ten minutes' concentration ceased. Fortunately we suffered no losses.

The companies moved rapidly to their assigned places. After the exertions of the day we slept well on the grassy field, wrapped in overcoats and shelter halves and in spite of a cold rain which started in shortly after our arrival.

II: Attack Against The Ridge Road Salient, August 9, 1917

A renewed surprise artillery concentration woke us abruptly before daybreak. Lieutenant Hausser, my adjutant, and I had bivouacked just above a small hollow where some shells burst alongside the pack animals tethered there. The latter broke loose and stampeded over us and out into the night. Shell after shell struck round about us, several just missing us by a hair's breadth. We waited until the fire began to subside before daring to make the short dash to the hollow which offered us better shelter.

The hostile fire soon ceased but this time several men had been wounded by shell fragments and Dr. Lenz had to take care of them. At daybreak I made my way to the battalion command post and, with hot coffee, restored myself from the night's scares and alarms. Toward 0500 we were ordered to move up the south slope of the Ungureana on a level with the 18th Bavarian Reserve Regiment and to continue the attack.

Under strong harassing fire we crossed the west slope of the Ungureana by moving through communication trenches and by dashing from crater to crater, and we felt relieved on reaching the less dangerous wooded southwestern slope of the mountain. On arrival I was ordered to take the 1st and 2d Companies and drive the enemy from the small wooded plateau half a mile south of the Ungureana summit.

First I established contact with the right of the 18th Bavarian Reserve

Infantry Regiment, which had dug in about a hundred yards up the slope during the preceding evening. Unfortunately I was unable to obtain information as to the location of the Rumanian positions for no reconnaissance had been made in the direction of the small plateau. For the first time I was able to examine the terrain over which I was to advance and I also checked the map thoroughly. A deep ravine lay between us and the plateau and both were covered with trees and dense underbrush.

I sent a sergeant with the men and a telephone detail out to locate the enemy dispositions and within fifteen minutes had the report that the strong position on the plateau had been abandoned by the enemy. On receipt of this information, I immediately pushed both companies forward by following the telephone wires in single file and seized the abandoned position and prepared it for all-around defense. I had to bear in mind that hostile forces coming from any direction might want to reoccupy the well-built installations. When I reported to Major Sprösser, scarcely thirty minutes had elapsed since the assignment of the mission.

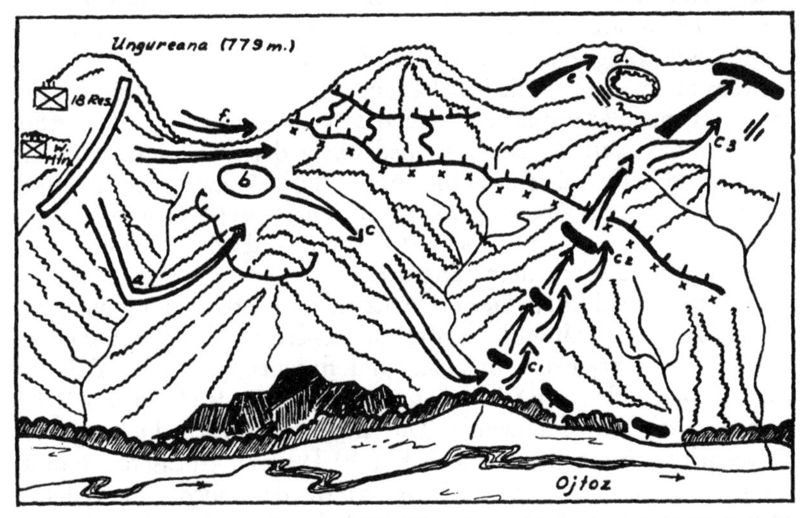

Sketch 21: Situation on August 9, 1917. View from the south. (a) Seizure of the plateau. (b) Noon rest. (c) Afternoon attack. (d) Evening position (e) Enemy counterattack. (f) Attack of 18th Bavarian Infantry and the Württemberg Mountain Battalion.

The chief activity of the forenoon consisted of reconnaissance of the almost roadless and wooded region toward the south (Ojtoz Valley) and east. Two prisoners were brought in during this time and we were relieved at noon by Honved (Hungarian) infantry. On battalion order, my detachment, which had been reinforced by the 3d Company, moved off through the woods and to the north in the direction of the ridge which lay a quarter of a mile southeast of the Ungureana. We used the same security measures as before (scout squad with telephone detail). On arrival we established ourselves with all-around security, for we had no direct contact on either flank, and I wished to avoid unpleasant surprises. Our enemy information was to the effect that he occupied very strong positions on the main ridge about half a mile east and northeast of the Ungureana. Following a short artillery preparation these positions were to be assaulted at 1500 and the enemy driven back beyond the bend in the ridge road, which was about a mile east of the Ungureana. The 18th Bavarian Reserve Infantry Regiment was to attack along the ridge line with the Württemberg Mountain Infantry Battalion just south of it. My unit was assigned to lead the attack.

While the companies rested and ate in the deep gullies to the west, I sent out several scout squads, each equipped with a telephone, toward the positions to be attacked in the afternoon. Technical Sergeant Pfeiffer and two men went off as the southernmost reconnaissance squad with the mission of finding out the location and strength of enemy garrisons on the ridge running south from the bend in the ridge road.

From the type of hostile installations on the plateau I concluded that the enemy had had insufficient time to develop a well-integrated position on the slopes farther east. It seemed likely that the positions on the heights and in the valley were the only well-developed ones, while those on the slopes were poor and disconnected. These last positions constituted the weak spot in the hostile defensive system and offered high rewards and quick success for enterprising troops.

The scout squad sent to the north encountered wired-in positions everywhere, but about half an hour after his departure Pfeiffer reported the capture of seventy-five Rumanians and five machine guns. This feat seemed incredible for we had not heard any sounds of firing. Pfeiffer soon telephoned the following: "The enemy was surprised resting without security elements in a ravine six hundred yards northeast of our unit's camping place. We discovered him as we descended, attacked him silently with two riflemen, and ordered the Rumanians to surrender. Since the Rumanians had laid their weapons to one side, they were defenseless and had to let themselves be captured."

I reported Pfeiffer's success to Major Sprösser and suggested that I take my units and break through the nonintegrated enemy positions on the northern slope at the same time as the frontal attack was being launched against the crest. If my attack was successful its gains could be exploited by seizing the ridge at the road bend, thus putting us in the rear of the strong enemy position east of the Ungureana and obliging him to evacuate his defensive system. Major Sprösser passed the proposal on to brigade and shortly thereafter I was ordered to carry out the proposed attack against the positions on the slope with the 2d and 3d Companies. Unfortunately I was not given any heavy machine guns.

Soon the unit marched silently down Pfeiffer's telephone line with his squad acting as advance guard. He had failed to locate any other enemy forces. We descended toward the valley and passed through a heavy forest of deciduous trees and thick underbrush. The slope was steep and I was obliged to follow Pfeiffer who led us down into the Ojtoz valley at a sacrifice of twelve hundred feet of elevation.

We were barely a hundred yards from the Ojtoz valley road when I caught up with Pfeiffer and ordered him to start climbing toward the salient in a northeasterly direction. Lieutenant Hausser, some runners and I went along at the head of the main body. Soon it became apparent that something was wrong, and I hurried forward. In a less dense part of the forest Pfeiffer pointed to some Rumanian sentries about two hundred yards away behind whom we could see the Rumanians' positions. The enemy was directing his attention to the open terrain on both sides of the valley road. We left them undisturbed and climbed up a narrow path leading through the thickly wooded steep west slope in the direction of the salient. It was quite obvious that we would run into the Rumanian positions during our climb, and I therefore ordered the advance guard to take cover as soon as contact with the enemy had been made and to protect the advance of the remainder of the unit. The advance guard was prohibited from opening fire unless attacked by the enemy. My idea was to deceive the Rumanians and let them believe that they had run into a reconnaissance detachment thereby gaining time to complete the ascent and prepare for the attack. By following these precautions I hoped to surprise the Rumanians.

Five hundred feet above the floor of the valley the advance guard was fired on from a position farther up the slope and, as per order, took cover without returning the fire. I quickly disposed the unit for attack with the 3d Company on the right and the 2d Company on the left. The thick underbrush made it possible to complete our preparations unbeknownst to the enemy. My attack order was:

"Second Company attacks astride the narrow footpath. The attack is a feint and must deceive the enemy and pin him down by means of rifle fire and hand grenades. Direction of attack is up the west side of the slope. Simultaneously the 3d Company envelops the hostile right. I will be with the 3d Company."

Some Rumanian reconnaissance detachments found their way into our assembly area and forced us to action before we had completed our preparations. They were repulsed and I immediately ordered the 2d Company to attack. The company encountered an occupied position 150 feet up the slope. During the ensuing fire and the hand-grenade battle, the 3d Company and I climbed some three hundred feet to the east passing through thick brush and reached the enemy flank without meeting any opposition. The enemy was in platoon strength and his attention was focused on the frontal fire fight. Our attack forced him to evacuate his position and retire up the slope. We were unable to pursue because of the dense forest terrain, the limited visibility, and the fact that a further advance would have brought us into the 2d Company's field of fire. Therefore I halted the 3d Company.

The 2d Company continued to press the retreating enemy repeating its former tactics wherever it met increased resistance. The 3d Company did likewise, and the retreating foe scarcely had time to halt and turn before he was driven to ground by the rifle fire and hand grenades of the 2d Company. These renewed outbursts were signals to the 3d Company to start another envelopment on the right. This type of combat under a burning August sun called for tremendous exertions on the part of the troops who had to contend with their heavy packs as well as with the steep slope. Several men collapsed from exhaustion.

We drove the enemy from five successive positions each one stronger than its predecessor until Lieutenant Hausser and I together with ten or twelve men were the only ones left in pursuit of the enemy. Steady shooting, shouting and hand grenades thrown to one side so that we might avoid their fragments as we charged forward kept the Rumanians on the run as they retreated through the undergrowth. In this way we succeeded in driving them back through a developed and apparently continuous position secured by obstacles, and prevented them from making a stand.

The woods beyond the position were less dense and the hillside while still leading upwards became less steep. We reached a forest clearing bordered on the right by long grassy slopes across which we saw two enemy companies retreating in a northeasterly direction towards the crest of the ridge. Over on the right a Rumanian mountain battery with its pack animals was displacing to the rear, trying to reach safety quickly. We opened fire rapidly from the thickets on the retreating enemy, who, fortunately, was not able to estimate our numbers. When the enemy had disappeared in the nearby woods and in the folds of the terrain, I ordered Lieutenant Hausser to continue the pursuit with all the available men.

When our mountain soldiers left the edge of the woods, a Rumanian mountain battery opened with canister from the left front in the northwest corner of the clearing about a quarter of a mile away. We took cover behind large beech trees. Shortly afterward the first of the 2d and 3d Companies came gasping breathlessly up the slope and I moved them to the right into a hollow which offered cover.

We were only about half a mile from our attack objective, the crest-line near the salient. The enemy's precipitous retreat called for a continuation of the attack regardless of the troops' condition. Sounds of heavy combat had been coming from the Ungureana for some time. The attack of the Bavarians and the other units of the Württemberg Mountain Battalion seemed to be making progress.

Our further advance toward the crest was barred by rifle and machine-gun fire. Even these few moments of respite had given the hostile leaders an opportunity to get their troops in hand and to form a new front. I was handicapped by not having a single machine gun in either company. By taking skillful advantage of the smallest irregularities of the terrain, we succeeded in moving closer and closer to the crest and to the enemy, who seemed to be well aware of the importance of his position. Anyone who showed himself drew an immediate burst of rifle and machine-gun fire. In this manner Technical Sergeant Büttler received an abdominal wound while observing close beside me.

Twilight began to favor our progress. Shortly before the fall of darkness the Rommel unit occupied the heights just west of the Rumanian crest position, which hitherto had given us so much trouble. Elements of my outfit began to entrench themselves in a small saddle seventy yards from the Rumanian rifle muzzles but in defilade from them. These trenches faced east. Other elements secured the adjacent woods on the west where they had the enemy to the north and west.

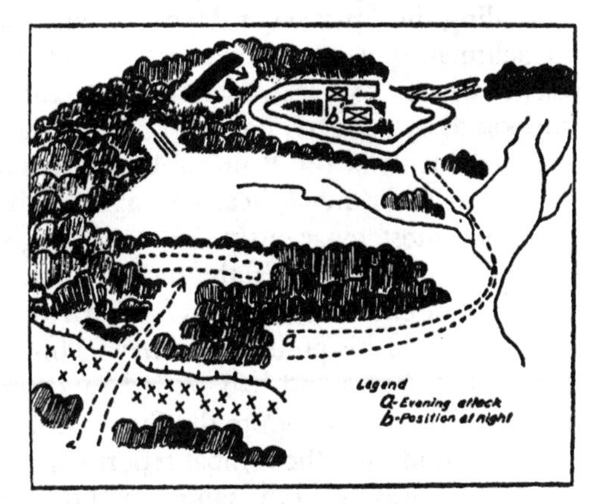

Sketch 22: Evening attack of August 9, 1917. View from the south.

Of course the Rumanian counterattacks tried to drive us from the height, but the lively carbine fire forced the attackers back to their starting positions. Since we had pushed a wedge across the ridge road, contact between the Rumanians in positions east and west of us had been broken. The wire line to the battalion, laid with so much effort during the advance and fighting had been cut and I was obliged to use pyrotechnic signals to announce our arrival at the objective to the battalion.

I made my dispositions in the dark and the unit dug in with all-around security, for we had to be prepared for counterattacks from all sides. I kept a platoon at my disposal in the oak wood close to my command post. We pushed combat outposts forward wherever the situation permitted.

We had no contact with the battalion. Apparently the frontal attack in the afternoon failed to achieve the desired result. Between the salient (we were about 550 yards east of it) and Ungureana, lively fighting continued to rage. Consequently we were about eleven hundred yards behind the hostile front.

In a pup tent I dictated my combat report to Lieutenant Hausser by the glow of a flashlight. Lights could not be shown anywhere without drawing immediate fire. Meanwhile the mountain soldiers performed an especially valiant deed. Lance Corporal Schummacher (2d Company) and a comrade carried the seriously wounded Technical Sergeant Büttler in a shelter half down to the Ojtoz valley (1100 feet difference in elevation). From there they carried their sergeant during the night to Sosmezö to a doctor who operated immediately and thus saved his life. In the dark night and considering the difficulties of the terrain and the length of the trip (eight miles as the crow flies) this was a tremendous accomplishment, a splendid example of soldierly fidelity.

Before the report was finished, I was relieved of the heavy worry regarding the situation at daybreak on August 10, for a reconnaissance detachment sent out in a westerly direction had made contact with elements of the 18th Bavarian Reserve Infantry Regiment. The latter, supported by artillery, had attacked frontally in the afternoon with the other elements of the Württemberg Mountain Battalion, but had been unable to make much headway against the enemy, who defended his positions most tenaciously. Then, through the noise of fighting and later by the light signals, the success of the attack by the Rommel unit had become clear to friend and foe. To avoid being cut off, the Rumanian had evacuated his positions between Ungureana and the salient under cover of darkness and had retired to a northeasterly direction toward the slopes leading down to the Slanic valley.

Before midnight the combat report was sent by runner to the battalion on the Ungureana. At the same time I ordered a new wire line to be laid. The night was cool, and I was so cold in my sweat-soaked clothes that I got up at 0200 and moved about to keep warm.

With Lieutenant Hausser I went to the front line and reconnoitered the hostile position, which lay opposite us to the east on a small wooded height (in the so-called oak wood) about ninety yards away.

Since I had forbidden unnecessary shooting because of the supply difficulties, the enemy was most uncautious. His sentries marched their posts as if under the most peaceful conditions and were most conspicuous against the eastern horizon, which was becoming lighter. It would have been simple to shoot them but I wished to defer this to a later time. At daylight we were able to determine that the Rumanians were occupying a continuous position to the east of us which began at the Petrei and ran almost due north through the woods.

Observations: The fire attack of the Rumanian artillery in the night of August 8 and 9 in the area where the Rommel unit lay in reserve caused a few losses. These losses would have been reduced had the troops dug in.

On August 9 combat reconnaissance by scout squads behind which telephone connection was laid proved excellent in the wood-covered mountains. I could call the scout squads at any time during the advance, get information in a few minutes, could give new orders or recall a part of the squads, or could by moving along the telephone line of the successful scout squad, quickly advance and occupy the position with my main body. The runner system, usually time-consuming in the mountains, was avoided. A preliminary condition, to be sure, was an abundant supply of telephone equipment.

In the difficult attack in the forest up the steep slope, the enemy, located in a higher position, was deceived as to our main attack by a lively fire, shouting and hand grenades, and was induced to dispose his reserves incorrectly. The thrust by the 3d Company against the flank and rear then led to a quick success. In the same way five such positions were taken one after another, though the final garrison was two companies strong. The attacks followed each other so quickly that the enemy had no time for regrouping.

In spite of the enemy's superiority in number and armament, the Rumanians had numerous machine guns and mountain guns at their disposal,—the Rommel unit, by taking advantage of the smallest irregularity of the terrain, succeeded in capturing and defending the crest of the heights eleven hundred yards behind the hostile front. The enemy was thus compelled to vacate his positions opposite the 18th Reserve Infantry Regiment and the Württemberg Mountain Battalion during the night.

After a successful attack the Rommel unit dug in quickly with all-around security. Without having dug in it would have suffered heavy losses from hostile fire and from the enemy counterattack. Our losses were 2 dead, 5 severely wounded, 10 slightly wounded.

III: Attack of August 10, 1917

Toward 0600 on August 10 telephone communication had been established with the battalion. Through the administrative officer I learned that Major Sprösser had received my combat report and had marched toward the salient with all his units.

Toward 0700 Major Sprösser arrived with the other companies of the Württemberg Mountain Battalion and he gave fullest praise to the Rommel unit for its conclusive success on August 9.

I oriented myself as to the situation toward the east in front of the unit. There the Rumanian sentries behaved incautiously even in broad daylight. In fact, some units of the Rumanian garrison were sunning themselves hard by the positions dug during the night between the Petrei Hill and the oak wood. Things were quite different with us. The sentries and garrison of the Rommel unit were well concealed and had strict orders not to let themselves be seen anywhere and to shoot only in case of a hostile attack.

The hostile positions stretched from the bare west slopes of the Petrei (693) (2286) along the ridge rising toward the oak wood; the ridge had only a few clumps of bushes. The oak wood itself seemed to be strongly fortified. It commanded the area toward the south, west, and north. North of the oak wood the enemy positions extended valleywards through the undergrowth toward the deep gorge of the Slanic. The positions consisted of individual nests and larger strongpoints, all mutually supporting and dominated the bare slopes to their front.

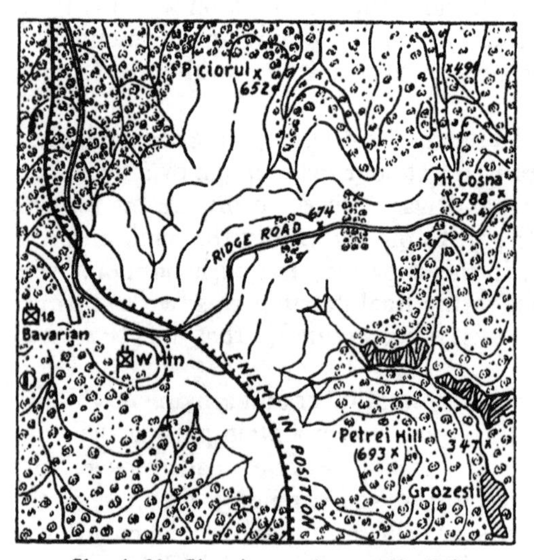

Sketch 23: Situation on August 10, 1917.

According to brigade orders, which arrived shortly after 0700, the mountain battalion was to continue the attack and seize the fork in the road four hundred yards west of Hill 674 (2223). Once again the enemy had to be driven from his positions. This attack was to be made without artillery support for our guns had insufficient time to displace forward. Major Sprösser detailed me to prepare and execute this maneuver and gave me the 1st, 3d, and 6th Mountain Companies as well as the 2d and 3d Machine-Gun Companies. This gave me command of a sizable force.

My plan of attack was to strike the unsuspecting enemy suddenly with machine-gun fire toward noon, force the hostile garrisons located in the area four hundred yards south to three hundred yards north of the oak wood to take cover, pin them down and, at the same time, break through in the region of the oak woods with some of my units, roll the enemy back close to the left and right of the oak wood and block him off. With these operations accomplished, I planned to take my main force and, in a single push, break through and fight my way to Hill 674 (2223).

The preparations were tiresome and time-consuming. During the forenoon I personally concealed ten heavy machine guns, moving them to their positions over a wide detour, so as to avoid hostile observation. Some were emplaced on the wooded crest of the heights close behind our forward line and the remainder were in the rills and folds of the south slope. I assigned targets to each gun and planned the fire schedules to be followed before, during, and after the attack. I set the opening of fire for 1200 and designated the platoon located nearest the salient as the base platoon.

Sketch 24: Fire plan for the attack on August 10, 1917. View from the south.

The remaining units of the Rommel detachment finished their preparations toward 1100. I selected the south edge of the oak woods as the breakthrough site. The depression ninety yards southwest of the oak woods was being filled silently with assault troops, namely: the 3d, 1st, and 6th Companies and a heavy machine-gun company. I issued orders and instructions to the assault team (3d Company), to the elements of the 3d Company which were to make the feint, and to my main attack force.

Mail arrived ten minutes before the attack and was quickly distributed.

Punctually at 1200 I gave the base machine-gun platoon the prearranged signal for opening fire. A few seconds later all ten heavy machine guns were in action. There was plenty of cover in the woods. In order to mislead the enemy and to cause hasty commitments on his part, the left-flank platoon of the 3d Company shouted as loudly as possible simultaneously with the opening of machine-gun fire and threw numerous hand grenades into the northwest corner of the oak wood. All this was done from cover in order to keep losses to a minimum. The Rumanians were not slow in answering our fire.

The 3d Company assault team covered the hundred yards to the southwest corner of the woods amid a deafening roar and was partially concealed by the smoky haze which arose from the explosion of many hand grenades. The heavy machine-gun companies had given the enemy a good work-out, and I ordered them to shift their fire to the right and left and make a narrow fire-free path for the assault team which moved silently forward firmly determined to turn in a complete job. My staff and I followed close behind the assault team and the remainder of the 3d Company with a heavy machine-gun platoon was right behind us. There were banging and shooting in all directions.

About two minutes had elapsed since we started firing and our ten heavy machine guns continued to hammer away and the uncontrolled noise of battle raged to the left of the road. The assault team forced its way into the oak woods and found it had to clear the enemy trenches, a job which our mountain infantry made quick work of. Whenever progress in the trenches was denied them, then they left their covered approach and enveloped the local strong point. These local turning movements received excellent support from our machine-gun platoons located in the woods who pinned the enemy down while our assault teams maneuvered. One of my combat orderlies killed a Rumanian who was aiming at me from a distance of about fifteen yards.

We were scarcely in possession of the enemy positions in the oak woods when we were hit by a strong counterattack coming from the northeast. None of our heavy machine guns was emplaced and irregularities of the terrain prevented the rearward guns from engaging this new enemy. The enemy was soon within hand-grenade distance and a tough grenade and carbine fight ensued in which the staff used its weapons as well as the men in the line. We held our ground in spite of an enemy superior in numbers, and the entry into combat of a heavy machine gun platoon turned the tide in our favor and allowed me to go back to my job of commanding the unit.

Elements of the 3d Company and a heavy machine-gun platoon secured the section of woods in our possession to the north and south. I assigned my remaining forces (1st and 6th Companies, as well as the elements of both machine-gun companies made available by our successful breakthrough) the mission of breaking through along the ridge in the direction of Hill 674 (2223). While some heavy machine guns pinned the enemy down in his positions on both sides of the oak woods other units blocked the shoulders of the breach in the hostile position thus allowing the main body to storm the ridge regardless of strong fire on all sides. Hill 674 (2223) was our only objective and we advanced in column of companies led by 1st Company, whose leading elements soon reached a small knoll a quarter of a mile west of Hill 674 (2223) without meeting any resistance. I was close behind them and was just crossing a small depression when I was forced to hit the dirt by a burst of machine-gun fire coming from the right. The bullets dug small holes in the greensward and we estimated their source to be a slope some nine hundred yards south of Hill 674 (2223), that is more than thirteen hundred yards away from us. I had had fair cover in a small fold in the ground and I intended to dash on as soon as the machine-gun fire stopped when I was suddenly shot in the forearm from the rear and the blood spurted out. Looking around, I discovered a detachment of Rumanians firing on me and a few men of the 1st Company from some bushes about ninety yards behind us. In order to get out of this dangerous field of fire I made a zigzag dash to the knoll in front of us where some elements of the 1st Company had to defend themselves for about ten minutes until the Rumanians to the west had been taken care of in hand-to-hand fighting by the men following us. The French officer commanding the Rumanian unit kept shouting "Kill the German dogs!"

Farther back violent fighting had also developed. The Rumanians had recovered from their initial fright and were trying to recapture their lost sectors by means of counterattacks with local reserves. The decision was ours thanks to the incomparable bravery of all the mountain riflemen and the energy of the officers.

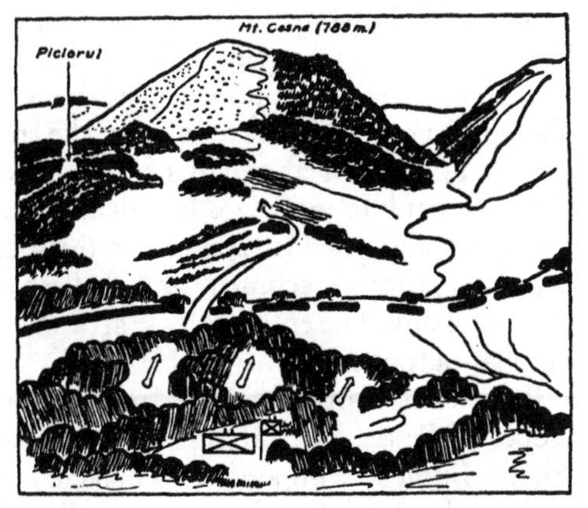

Sketch 25: Mount Cosna. View from the west.

The 1st and 6th Companies took Hill 674 (2223) without encountering further resistance. Meanwhile my arm was bandaged by Dr. Lenz. Then I ordered my unit to occupy the captured territory and to reorganize. The order was:

"The 6th Company, reinforced by Aldinger's heavy machine-gun platoon, on Hill 674. All other units at my disposal in the broad hollow just north of the road four hundred yards west of Hill 674."

In spite of severe pain and exhaustion through loss of blood, I did not give up command of the unit. Major Sprösser was informed of our success by telephone.

About this time a long column was seen marching toward us on the ridge road from the direction of Mount Cosna. So we organized the defense, and the spade came into its own. I urgently requested artillery fire on approaching hostile forces but this request could not be fulfilled as all artillery units were displacing forward. The enemy drew nearer unhindered.

Captain Gössler arrived with the remaining companies of the Württemberg Mountain Battalion and we split the command. The Rommel detachment consisted of the 5th and 6th Companies and Aldinger's machine-gun platoon as front-line garrison and the 2d and 3d Companies and the 3d Machine-Gun Company as second-line garrison. Gössler was given the 1st and 4th Companies as well as the 1st Machine-Gun Company. His detachment was dug in some three hundred yards west of Hill 674 (2223) just south of the ridge road. Contrary to expectations, the Rumanian infantry approaching from the direction of Mount Cosna did not counterattack. They merely contented themselves with feeling out our position with strong reconnaissance detachments which were easily repulsed. Following this, the Rumanians occupied the ridge opposite the 5th and 6th Companies. Their position was half a mile away and was some twenty-two hundred yards long. Under these circumstances there was no need for us to increase the front line garrison. The 5th and 6th Companies together had a front of about seven hundred yards with the open flank refused. Gössler's detachment was in contact with the 6th Company and provided security on the southern slope while the remaining units of my detachment provided security for the north flank of the 5th Company. The entire defense area was further secured by a system of combat outposts in considerable depth.

The Rumanians withdrew from the line extending from the west slopes of Petrei Hill through the oak woods to the west bank of the Slanic. But it was impossible to make contact with our neighbors to the right and left. Violent Rumanian artillery fire began and soon destroyed the wire connections, denied all movement to runners, and cut up the terrain on both sides of the ridge road between the oak woods and Hill 674. The telephone connections with the 5th and 6th Companies were repaired repeatedly, a difficult and dangerous task for the wire details. The fire persisted during the entire afternoon with undiminished violence. Fortunately, the companies up forward and the reserve areas were not seriously inconvenienced. In the late afternoon the Austrian artillery made itself felt. Among other things, a 30.5cm shell struck in the midst of a group of men (as it later turned out, a group of Rumanian and French officers) on Mount Cosna's summit. Fortunately, my detachment losses during the attack and the subsequent artillery bombardment were very low. During the bombardment, I prepared my report in my command post located in the steep slope four hundred yards west of Hill 674 (2223). The hostile artillery fire did not stop until dark when our pack train came up with rations and ammunition.

I was exhausted by loss of blood, and the tightly bandaged arm and overcoat thrown over my shoulders hampered every movement. I was considering giving up the command but the detachment's difficult position prompted me to remain at my post for the time being.

Additional troops were put under Major Sprösser's command. His command post was in the oak woods twenty-two hundred yards southwest of Hill 674 (2223). There, too, were the reserves of the Sprösser group (units of the 18th Bavarian Reserve Infantry Regiment) and the observation posts of the artillery liaison officers.

Night fell.

Observations: The attack by the Rommel detachment on August 10, 1917 against the commanding, fortified Rumanian position had to be carried out without artillery or mortar support. Only heavy machine guns were available to support the attack. The attack was successful and cost little in way of casualties because: first, we had prepared a heavy concentration of machine-gun fire on that point in the hostile position where the 3d Company's assault team was to break through; and second, we succeeded in pinning the enemy down with machine-gun fire both during and after the initial assault.

On August 10 the Rumanians did not make the mistake of the preceding day, when they neglected the position on the slope. A breakthrough into the hostile position half way up the slope would have promised little success on August 10, since the terrain was open, and such an attack could easily be blocked off by machine-gun fire from the heights round about. The enemy had to be tackled along the ridge itself.

Battle reconnaissance: Sharp observation of the hostile territory yielded excellent results during the night of August 10 and in the first hours of the morning. The forward hostile installations and the behavior of the garrisons were accurately ascertained. Scout squads were not sent out by us, in order not to arouse the enemy and make him curious as to our attack preparations. The enemy, however, committed the great mistake of not surveying the terrain in front of his position, and in fact behaved in a most unwarlike manner (visible sentries, garrisons outside of their shelters). Thus our surprise attack struck him like a thunderbolt.

The assault team of the 3d Company had a path to the oak woods prepared by several heavy machine guns which covered the enemy in the oak woods with combined fire from positions two hundred yards west of the breakthrough site and then shifted their fire to right and left, so that the advancing squad of the 3d Company was not endangered. In the further course of the attack the same heavy machine guns admirably supported the rolling up of the hostile positions by laying their fire close in front of their own assault teams.

The feint one hundred yards to the left of the breakthrough point delivered from complete concealment with hand grenades and shouting was to draw the defensive fire of the enemy in the oak woods in a false direction and to bring about the premature commitment of reserves. It fully achieved its purpose in helping the assault team forward, without producing any losses.

To be sure, the enemy quickly and skillfully delivered a counter thrust from the northeast against our breakthrough in the oak woods, but the superior fighting ability of the mountain riflemen also proved itself in the defense.

The Rumanians had occupied the crest of the heights to the rear of the continuous position with reserves but the latter were, for the most part, not prepared to react against our surprise breakthrough and were overrun in their dugouts. Wherever they took up the defense or counterattacked, they were quickly overwhelmed by the greater strength of the mountain soldiers, for five companies advanced through the breach to be followed by the Gössler detachment and four more companies. Thus the surprise attack had the necessary power.

After seizing the objective we went over to the defense. The companies of the front line dug in with good concealment. The open flanks on the north and south were secured by combat outposts from the reserve company. It was not advisable to send out scout squads to greater distances. They might easily be shot or captured by the garrisons of rearward Rumanian positions. On the other hand, the hostile territory was most thoroughly studied from the various OPs. Shortly after reaching the objective, our troops vacated the ridge between the oak woods and Hill 674. They had dug in laterally on the irregularities of the terrain. The very heavy hostile artillery fire in the afternoon did little harm.

The attack of the Rommel detachment along the ridge forced the enemy to evacuate his breached position in the afternoon and to withdraw to a new position.

The hostile command was not very active, limiting itself to nothing more than defense and not daring to launch a resolute counterattack, although numerous reserves and strong artillery were at hand, and the terrain to the north would have been, like that in the south, most favorable for a counterattack.

IV: The Capture of Mount Cosna, August 11, 1917

The front remained quiet and we were not even bothered by Rumanian scout squads. Toward 2200 Major Sprösser informed me that the brigade had ordered an attack with artillery support against Mount Cosna for 1100 the next day, and was asking for suggestions.

Judging from the terrain, an attack from the west and northwest seemed most promising to me, for here the highest parts of the mountain ridge were not wooded and artillery and heavy machine-gun support would be easily secured. Moreover, the numerous folds of the terrain north of the ridge road offered good avenues of approach for the attacking troops.

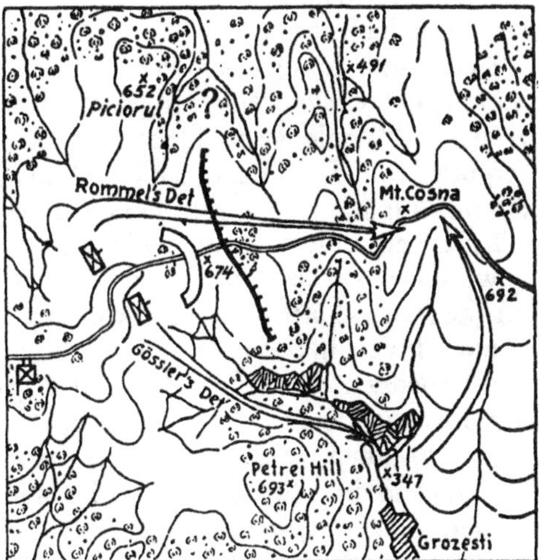

Sketch 26: Attack plan for August 11, 1917.

Major Sprösser then requested me to remain with him an additional day in spite of my wound, and to take over the command of the group attacking from the west and northwest. The 2d, 3d, 5th and 6th Mountain Companies, the 3d Machine-Gun Company, and the 1st Machine-Gun Company of the 11th Reserve Infantry Regiment were assigned to me. At the same time the southern attack group under Captain Gössler (1st and 4th Mountain Companies, 1st Machine-Gun Company, 2d and 3 Battalions of the 18th Bavarian Reserve Infantry Regiment) was to attack Mount Cosna from the south or southwest via Hills 647 (2136) and Hill 692 (2290). The new and difficult task was most attractive so I remained with the outfit.

I got little sleep during the night because my wound smarted and my nerves were on edge as a result of the day's activities not to mention my preoccupation with the morrow's work. Before daybreak I woke Lieutenant Hausser and we went forward to the 5th and 6th Companies and, in the early morning light, studied the terrain and prepared our attack plans.

The enemy positions were astride the ridge road on the next ridge half a mile to the east of our forward positions. His sentries were hidden behind trees or in the undergrowth. North of the road we located a fairly compact skirmish line in recently dug positions. Elements of the garrison stood in groups. Neither side disturbed the quiet of daybreak with shots. Our positions were well concealed and scarcely perceptible to the enemy.

The avenues of approach were less favorable than I had thought. Bare grassy slopes in front and to the south offered no protection against hostile fire. The terrain seven to nine hundred yards north of the ridge road appeared more favorable. On the grassy slopes of the ridge leading to the Piciorul numerous fairly large and dense clumps of bushes were scattered about. The Piciorul (652) (2200) located a mile north of the ridge road on the flank of the 5th Company was covered with large deciduous trees.

Sharp and dominating, the Mount Cosna summit loomed on the horizon in the rays of the rising sun. It was the objective for the attack on August 11. Would we be able to do it? We had to! My wounded arm was forgotten for I had six companies to lead against the enemy. I went about the difficult and responsible work with confidence and new strength.

I planned to use the companies already in position to pin the enemy down starting at 0800, to mislead him and prevent him from reconnoitering the ravines northwest of his positions. During the morning I intended to move my main body south of the Piciorul to within attacking distance of the enemy north of the ridge road. In executing this movement maximum use would be made of all available cover. Once in position I expected to attack with artillery support at 1100 hoping to breech the position and drive through to the Cosna in a single movement. The units located on Hill 674 were to launch a frontal attack coincidental with ours.

The 5th and 6th Companies, with Aldinger's machine-gun platoon, were given to Lieutenant Jung, whom I instructed through Lieutenant Hausser, as to my plan and the tasks of his formation in the attack on Mount Cosna. I left Lieutenant Hausser with Jung's detachment, in order to secure communication with the Sprösser group and cooperation from the artillery.

At 0600 I moved off to the north through dense shrubbery with the remaining four companies. Telephone wire with Jung's combat group was laid at the same time. After about seven hundred yards I turned the head of the column eastward and we approached the ridge between

Hill 674 and the Piciorul by climbing up a shallow draw. The ridge was sparsely covered with lone trees and clumps of bushes. Now and then we halted and studied the terrain, and I was amazed to see that the enemy had combat outposts along the entire ridge. The Rumanians had pushed combat outposts out in front of their new position. Neither the 5th Company, on whose left flank the outposts were located, nor the scout squads of the reserve companies had located these outposts.

Under these conditions a surprise attack from the northwest against the Rumanian main position seemed almost impossible. If I overran the hostile outposts, then the enemy in the main position east of Hill 674 would be alerted and my attack would no longer be a surprise which would materially reduce the prospects for a successful attack.

We halted concealed from hostile view. Thorough consideration of the terrain round about led me to decide to outwit the combat outposts in front of us. We retraced our steps and, after going a short distance, turned to the north and reached the dense zone of woods on the northwest slope of the Piciorul without encountering the enemy. Again we turned to the east and moved through the dense underbrush of the tall forest toward the Rumanian combat outposts.

I organized my own security in greater depth. Far in front an especially skillful technical sergeant of the 3d Company was scouting, and I directed him by means of arm signals and low calls. Upon my request, his platoon leader, Lieutenant Hummel, was carrying his heavy pack on his own shoulders. I marched a few yards behind the technical sergeant followed by the remaining ten men of the advance guard who marched at ten pace intervals. The four companies followed in single file 160 yards behind the advance guard. This distance was so arranged that when my signals halted the advance guard the companies could continue the march without giving telltale sounds. Naturally, absolute quiet prevailed in the whole detachment which was in a column about half a mile long. Each soldier avoided making the slightest noise. The troops knew that it was a matter of moving unobserved through the hostile combat outposts.

We halted and resumed the march on signal. By listening for some minutes we succeeded in determining the location of two Rumanian outposts. The hostile sentries talked, cleared their throats, coughed, and whistled, as we approached yard by yard. The hostile sentries were at 100- to 150-yard intervals but we could not see them because of the dense underbrush. I moved with the advance guard to the middle of a gap between two hostile sentries. We were on a level with them and held our breaths. The enemy to right and left did not diminish his conversation and I carefully moved the four companies through. At the same time telephone connection with Jung's combat group was being laid for this line also connected us with the command post of the Sprösser group. The adjacent enemy was most unobservant.

Always slipping through dense undergrowth we reached the north slope of the Piciorul in the rear of the Rumanian sentries and field outposts who were still observing the front to the west. Meanwhile, on the right and according to our plans, Jung had opened with rifle and machine-gun fire.

A very deep ravine still separated us from the main Rumanian position and we had to negotiate the obstacles unobserved. In descending we crossed several roads, but fortunately we encountered no Rumanians. Up on the right near Hill 674 Rumanian artillery was plastering Jung's position with heavy fire. The Rumanians apparently suspected preparations for attack there, and were taking measures to forestall it.

The climb up the steep slope in a blazing August sun with the heavy pack, (the heavy machine gunners carried loads of almost 110 pounds on their backs) was a terrific exertion. It was almost 1100 when we reached the lowest point of the ravine and began to climb the abrupt, rocky slope of the other side. We proceeded slowly as the terrain caused great difficulties. Our artillery opened its fire for effect at 1100 sharp. To be sure, it seemed rather weak to us, and did not strike in the region where we were to attack. The fire column of the 5th and 6th Companies increased and the enemy answered with artillery.

During this period we bent every effort in climbing the slope. My wounded arm hampered my climbing very much and my combat orderlies had to help me over the more difficult spots.

Our own fire for effect had ceased when, toward 1130, the technical sergeant of the 3d Company who was out ahead as scout was fired on in a light forest and, as instructed, quickly took cover without returning the fire. I ordered the advance guard to halt and secure the ascent of the companies who came up silently until they reached a narrow space on the protecting slope about 160 feet below the advance guard. While this was going on I got Jung on the telephone and told him I intended to attack in half an hour. I also tried to get in touch with Major Sprösser and ask for artillery support but the wire went dead. Apparently the Rumanian detachments on the Piciorul had discovered the wire and cut it.

That the connection with the Sprösser group, the artillery, and the Jung combat group should give out just before the decisive attack was most unpleasant. To restore communication seemed barely possible and would take hours of hard work. I had to accept this misfortune.

The location of the enemy positions which we were to attack could only be surmised. I believed it to be in the region where the scout had been shot at by the Rumanian sentries. The configuration of the slope and the growth of bushes and high ferns made it possible to assemble in a well concealed area within rushing distance of the enemy. Support of the attack by machine-gun fire from elevated positions was out of the question, nor could Jung cover our front with fire for we had no communication whatsoever with him, but I hoped he would act according to his instructions.

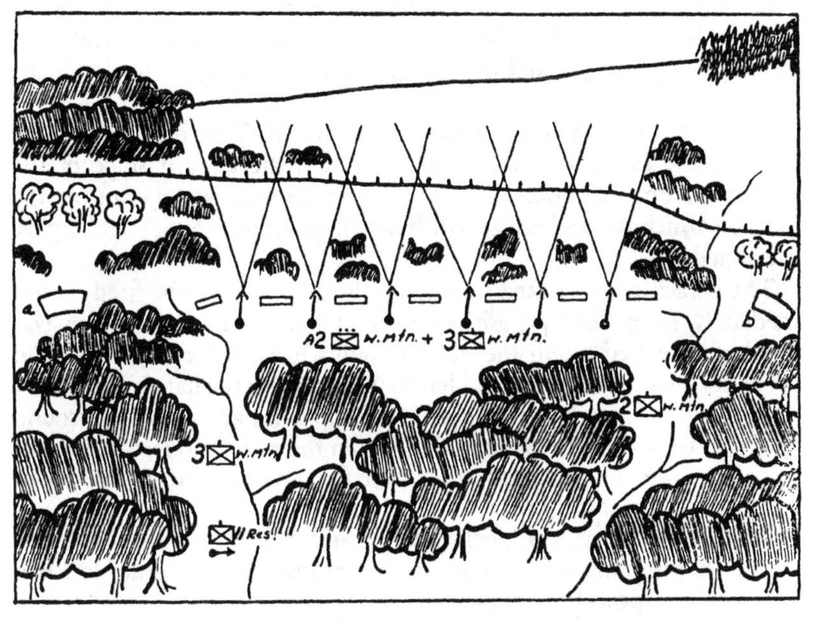

Sketch 27: Preparations for the attack, August 11, 1917. View from the west

I took one platoon of the 3d Company and Grau's machine-gun company and disposed them in the front line on a width of some one hundred yards. The 2d Company was echeloned to the right rear and the remaining two platoons of the 3d Company and the 1st Machine-Gun Company of the 11th Reserve Infantry Regiment were echeloned to the left rear. My attack order was:

"On my signal the forward line (1 platoon of the 3d Company and Grau's machine-gun company) creeps forward silently through the ferns toward the assumed position up on the slope. As soon as hostile sentries or the garrison open fire, Grau's machine-gun company combs the hostile position with continuous fire of all guns and stops on my signal after about thirty seconds. At this moment the platoon of the 3d Company and the other units of the detachment, which has to be kept closed up, break in on the hostile position without shouting. Individual squads block the shoulders of the breach immediately and the main body breaks through to the defensive zone of the enemy and seizes the ridge as initial objective and prepared to advance to the southeast. To deceive the enemy as to the place of breakthrough and to disperse his defensive fire, the sectors of the hostile position on both sides of the breakthrough point are to be engaged by hand-grenade squads."

All these preparations and discussions were carried out noiselessly within a hundred yards of the hostile sentries. Since I had left Lieutenant Hausser with the 5th and 6th Companies, I was obliged to do all the work myself.

We were ready a few minutes before noon. The Rumanians had done us the favor of not disturbing us. On the eastern slope of the Piciorul, Rumanian detachments of platoon size were crossing the path by which we had advanced. It was high time to attack and I gave the signal.

The detachment worked its way up the slope only to be fired on immediately from enemy positions near at hand. The enemy fire was quickly answered by all the machine guns in Grau's company. Hand grenades burst to the right and left as we lay ready to charge. The heavy machine-gun fire in front of us pinned the hostile garrison to the ground and only left the enemy firing wildly from the right and left. I gave the signal to stop the heavy machine-gun fire and the mountain troops stormed up the slope, broke into the hostile position without any real losses, took a few captives, blocked off the area, and then charged forward to the right into the defensive zone. Everything went with the clocklike precision of a peacetime maneuver.

Soon the bushes in front of us began to thin out and we advanced another hundred yards before lively machine-gun fire hampered our advance against a slope rising gently to the right. The fire coming from a wood located on the highest hill about six hundred yards away across a broad grassy surface increased in violence.

The platoon of the 3d Company and the heavy machine guns of Grau's company took up the fight, and the remainder of the 3d Company and the machine-gun company of the 11th Reserves spread themselves out to the left. The enemy on the edge of the wood was being reinforced and we were soon engaged with several dozen machine guns. There was no question of continuing the advance across the grassy unprotected area for, in our tired condition, we were having trouble holding our own.

Hostile reserves counterattacked from the woods with artillery support and made their main effort against our left. The mountain soldiers clung desperately to the ground. They did not want to give way and their rapid fire stopped the hostile counterattack.

More and more enemy machine guns began to hammer us and our losses began to mount at an alarming rate with the result that our predicament grew more perilous with each passing second. I was up in front on the right of the 3d Company. On my left Albrecht's heavy machine-gun platoon was engaged in most violent combat. The 2d Company was in reserve to the right rear in the bushes where it was protected from the hostile fire. Should I commit my reserve? Would its fire power turn the tide of battle in our favor? No! Should I order a withdrawal? No! For then our dead and wounded would have been left in enemy hands and we would have been driven from this position down into the ravine where the Rumanians would have annihilated us with ease. The situation seemed desperate, but we had to master it or . . . remain on the spot.

There were some clumps of bushes down the slope to our right. The idea occurred to me that we might use these to cover an advance against the enemy on the hill and I decided to commit my last reserves in a surprise blow against the left flank of the enemy who was pressing us so hard. This move could decide the issue.

I gave instructions to those near me and crawled back and in a few seconds the 2d Company and I were rushing impetuously to the south.

Sketch 28: Situation at Mount Cosna, August 11, 1917. View from the west.

It was a case of do or die. We overran a weak enemy in the clumps of bushes before he knew what had hit him and in no time we had gained more than a hundred yards. We turned eastwards and I hoped that the remainder of the detachment was continuing to resist.

I was just about to launch the attack against the hostile flank when elements of Jung's group appeared on the right rear of the 2d Company. Jung was continuing the execution of his mission of the morning and was about to attack the enemy astride the ridge road. His arrival decided the battle in our favor for the enemy had committed his entire forces against the 3d Company and the two machine-gun companies and he had nothing left to throw against the attack of three mountain companies against his flank and rear. The Rumanians hastily vacated the height leaving the greater part of their machine guns on the battlefield.

On the edge of the wood, seven hundred yards east of Hill 674, the exemplary brave Lieutenant Jung, a leader respected by his company received a fatal abdominal wound.

The 3d and 2d Companies, as well as elements of the machine-gun companies, continued to fire on the enemy as he swept back in complete disorder along the ridge road and through the broad hollow. At the same time I took the 5th and 6th Companies and pursued the enemy just south of the ridge road and across the highest part of the ridge. The other units of the Rommel detachment received orders by runners to follow by the same route as soon as possible.

While the 6th Company took possession of the knoll half a mile west of Mount Cosna summit, we called it Headquarters Knoll, the 5th Company was bagging more than two hundred prisoners in protected hostile positions west and south of the ridge road and was capturing several machine guns. A broad ravine still separated us from Mount Cosna itself.

Dense masses of Rumanians were retreating down the road leading down the western slope and they were soon being hit by fire from the 6th Company. Rumanian troops were standing on the summit of Mount Cosna and we began to receive lively machine-gun and rifle fire from there. During it, among others, my splendid adjutant Hausser received a chest wound.

Soon the companies arrived one after the other on Headquarters Knoll. They were completely exhausted. No wonder, for since 0600 they had been marching, climbing difficult terrain, or attacking. Rest had been an idle dream.

The enemy occupied prepared positions on the steep height of Mount Cosna and could not have been attacked with exhausted troops. My decision was to rest the men and reorganize the units before considering an attack against the Mount Cosna summit position. The 2d Company furnished the security details for our rest area and a reconnaissance detachment from the 6th Company with wire reconnoitered the avenues of approach into the Mount Cosna position. From Headquarters Knoll we saw Tirgul Ocna lying northeast of us in the valley. The airline distance was not more than three miles, and we could see that heavy rail movements were in process at the Tirgul Ocna railroad station.

Toward 1300 the staff of the Sprösser group arrived together with the group reserves (2d and 3d Battalions of the 18th Reserve Infantry Regiment) just west of Headquarters Knoll. Major Sprösser had followed the attack of the Rommel detachment from his command post in the oak woods, and thought that we had taken Mount Cosna in one rush.

At that time nothing was known of the activities of Gössler's detachment. I announced my intention of continuing the attack on the summit position in an hour's time and asked for fire support from Headquarters Knoll by the machine guns of one of the two Bavarian battalions. My intention was to repeat the successful maneuver of the morning and Major Sprösser gave his consent.

At the agreed time, units of the 2d Battalion of the 18th Bavarian Reserve Infantry Regiment opened fire on the hostile positions. At the same time I climbed down into the ravine to the east and a hundred yards north of Headquarters Knoll with the 6th, 3d, 2d, and 5th Companies, the 3d Machine-Gun Company, and the 1st Machine-Gun Company of the 11th Reserve Infantry Regiment. We followed the reconnaissance detachment's wire through thick underbrush and down an extraordinarily steep slope. Soon we were going up the opposite side and had caught up with the reconnaissance unit from the 6th Company. The hot noonday sun made the climb most strenuous and it required several hours to reach the top with my exhausted men.

With security precautions similar to those of the forenoon we felt our way nearer and nearer to the enemy and climbed up through light brush and small rills. The garrison of the summit meanwhile was engaged in a lively fire fight with the 2d Battalion of the 18th on Headquarters Knoll and the fire from both sides whistled by high over our heads.

It was apparent that a Rumanian combat outpost was some two hundred yards away from the Bavarians on Headquarters Knoll. Finally we reached a small hollow some eighty yards from the summit. The Bavarians had ceased firing on the hostile position sectors above in order not to endanger us and the enemy's fire likewise had ceased.

I prepared my detachment for the assault with extreme care with two rifle platoons and six heavy machine guns in the front line and two companies echeloned behind each flank. The attack preparations were identical with those of the morning: creeping up, steady fire from the heavy machine guns, hand grenades on the right and left for diversion, and then the final assault.

Preparations were still incomplete when we plainly heard carbine fire in a southwest direction. Those sounds came from Gössler's detachment so I immediately gave the signal for the attack. After a short but continuous burst of fire the mountain troops smashed through the summit position and, in a few minutes, swept the west slope of Mount Cosna clear of the enemy. The enemy was so surprised that he failed to offer serious resistance in any portion of the position and the summit was ours at slight cost in casualties. We had several dozen prisoners and a few machine guns as our bag of trophies, but the major portion of the garrison of the position escaped and fled precipitately down the eastern slopes of Mount Cosna. As we started in pursuit, very strong Rumanian machine-gun fire struck us on the bare eastern slopes. This came from positions lying six to seven hundred yards east of Mount Cosna summit on a ridge running through Height 692 (2300) from north to south. These positions were particularly well developed and protected by wide obstacles. Strong artillery and machine-gun support were required before we could think of crossing the ridge and going down the eastern slope. We had to be satisfied with possession of the peak from which we could see far out into the Rumanian countryside.

We soon had contact with the 1st Company (Gössler's detachment) which was climbing up the steep ridge from the south toward Mount Cosna summit (788) (2500). The Rommel detachment dug in with the 1st Company (which I attached to myself) on the sharp slope south of the ridge road. The 5th and 6th Companies were on the peak and north of the ridge road descending to the northwest. I split the machine-gun company of the 11th Reserve Infantry Regiment among the three companies in the front line and kept the 2d Company at my disposal behind the center. The 3d Company and 3d Machine-Gun Company were behind the left flank.

About an hour after capture of Mount Cosna, Major Sprösser came up with both Bavarian battalions. Concerning Gössler's detachment we learned that after capturing the Rumanian positions near Hill 647, it came upon very strong enemy forces which, supported by numerous hostile batteries, attacked in dense masses from the east. Gössler's detachment had to be withdrawn because of heavy losses and was to halt on the east slope of the rocky ravine leading to Cosna summit from the south. On the left, toward the Slanic valley, our neighbor, the Hungarian 70th Honved Division, was still several miles away and out of contact with us. During the evening hours we watched the artillery duel north of the Slanic valley from our summit and observed the attack movements of Rumanian infantry in the region of Height 772 (2550).

I made arrangements for the night. Among other things, scouts were to establish contact with Gössler's detachment. The various companies were instructed regarding their responsibilities. I was so exhausted that I was unable to prepare my combat report for the Sprösser group. Through my new adjutant Lieutenant Schuster, I made a verbal report regarding the course of the day's fighting.

In spite of fatigue I found little rest that night. An hour before midnight, numerous hand grenades burst in the position of the 6th Company. Shouting, rifle and machine-gun fire resounded. Without waiting for a report, I counterattacked with the 3d Company in the direction of the threatened place, but when we arrived the 6th Company was already master of the situation.

What had happened? Rumanian assault squads had surprised the company but were repulsed by the watchful soldiers. But during the attack some machine gunners of the Machine-Gun Company of the 11th Reserve Infantry Regiment were taken prisoner.

Observations: The plan of attack for August 11 was developed as a result of personal reconnaissance during the early morning hours. The normal attack astride the ridge road supported by heavy machine guns and artillery was rejected because of the open terrain. It would have been seen early by the enemy and would probably have been repulsed with heavy losses.

The Rumanians had learned something from the battles of the preceding days and had set up combat outposts to secure the main position. This was detected in plenty of time by sharp observation of the battlefield during the approach march.

Only with a unit accustomed to the strictest combat discipline could I dare to feel my way through the hostile combat outposts by day.

The time and space calculations of this type of flanking marches are most difficult in the mountains. Here the unexpected appearance of the enemy was in addition to the terrain difficulties.

Cooperation with the artillery groups did not materialize during the attack, because the telephone connection broke down at the decisive moment. The artillery here would have been able to give good support to the difficult attack by the Rommel detachment.

The very difficult situation after the successful breakthrough was ha: dled by means of the reserve company. The thrust in flank and rear of the superior enemy rapidly turned the tide in our favor. In this connection the "Attack Schedule" given to Jung's detachment ahead of time proved to be extremely valuable, for even Jung was no longer in contact with us.

The fleeing Rumanians were not only shot at, but units of the Rommel detachment were immediately dispatched in close pursuit which was soon halted by rearward hostile forces in commanding positions.

While the exhausted assault troops rested, a scout squad reconnoitered the avenues of approach into the summit positions on Mount Cosna. Again the telephone line proved most useful.

The breakthrough into the hostile position at noon, as well as the breakthrough into the summit position in the evening, took place without artillery or heavy machine-gun support from rearward positions. Only the machine guns located in the front line of the assault troops covered the breach with their fire. Again the fire of the hostile garrisons of the positions was diverted to hand-grenade squads. The losses in the breakthrough itself were extremely slight.

The garrisons of rearward Rumanian positions received the retreating troops both upon the breakthrough at noon and upon the capture of the Mount Cosna summit and halted our pursuit.

V: Combat On August 12, 1917

The moon rose shortly after midnight and the scout squads sent out to Gössler's detachment reported that this unit was located with its left flank about half a mile southeast of Mount Cosna summit. It had sustained heavy losses and urgently requested support for the enemy was six hundred yards away and occupied very strong positions.

At 0100 I went on reconnaissance with some of my officers in order to examine the terrain in front of the right half of our position. I wanted to close the gap between Gössler's detachment and my right flank with a company before daybreak, and I also wanted to move my own position forward to within attacking distance of the hostile positions east of Mount Cosna; but Major Sprösser did not agree to that. He ordered the two Bavarian battalions to break through the hostile positions northeast of Mount Cosna at dawn, while units of the mountain battalion under my command followed the Bavarians in second line, prepared to exploit a successful breakthrough as far as Nicoresti.

Even before daylight we began to receive heavy artillery fire from a northwesterly direction—that is, from the left rear. It came from the heights on the far side of the Slanic valley. Their fragmentation effect was slight, but the shells dug craters twenty to twenty-six feet in diameter and nearly ten feet deep in the soft loamy soil. Lumps of earth fell in an area a hundred yards in diameter. Sleeping was out of the question and we had to move whenever the hits began to come too close. The fire increased and other batteries to the east and north selected Mount Cosna as their target, with the result that things became most unpleasant and uncomfortable.

Shortly before daybreak, two Honved (Hungarian) battalions which had been attached to Major Sprösser, arrived on the summit. One of them deployed on arrival, passed through my detachment and without orders proceeded to attack the Rumanian positions east of us. It suffered heavy losses and increased the hostile artillery fire.

I was much relieved when I led my detachment, consisting of the 5th, 3d, and 2d Companies, the 3d Machine-Gun Company, a Honved rifle company and a Honved machine-gun company out of the endangered areas. The two Bavarian battalions had started out ahead of us in order to execute their mission of breaking through the Rumanian position northeast of Mount Cosna by daybreak. A successful breakthrough would open the road to the plains and hasten the collapse of the Rumanian Mountain front south and north of the Ojtoz valley.

We crossed the west slope of Mount Cosna in a long column about six hundred yards below the summit and we were often endangered by Rumanian shells of the varied calibers, which struck all about us in a most unpredictable fashion. Marching in the cool of the morning made us feel very cheerful. After a half hour's march through light undergrowth on a steep slope, we reached the ridge descending from Elevation 788 (2500) toward Elevation 491 (1620). Tall fir trees covered the steep northeast slope and below on the left were small sections of a continuous fir forest. Through the fire we got a bird's-eye view of the Rumanian positions northeast of Mount Cosna, which the two Bavarian battalions had to penetrate; they consisted of carefully developed trenches with continuous broad obstacles out in front. Numerous communication trenches led over the bare ridge to the wooded zone on the east slope; between us and the hostile position was a draw which widened toward the northwest and whose slopes were covered with scrub oak.

The hostile positions were not ours. From twelve hundred to sixteen hundred yards north of us, we saw units of the Bavarian battalions in the broad draw just in front of the Rumanian positions, engaged in a hard struggle with the positions' garrisons.

We passed a group of wounded from the 18th Reserve Infantry Regiment and heard that all was not well up forward. Their leading battalion came upon the hostile position suddenly and suffered heavy losses (about three hundred wounded) from small-arms fire, with the net result that the breakthrough into the hostile position failed.

I ordered my detachment to fall out and rest and telephoned Major

Sprösser that lines had been laid, and informed him regarding the situation north of Mount Cosna. In view of the Bavarian's failure, my opinion was that artillery support would be necessary if we were to take the strong and well-constructed positions northeast of Mount Cosna. Artillery support was promised for the attack, but no artillery observer was available and I offered to adjust the fire, since my present location was an excellent observation post.

We examined the possibility of getting down into the hollow without being observed but could find no concealed avenue of approach, for the trees were too well spaced. I adjusted my first artillery fire at 1130 and at that time my detachment began its descent in column of files with twenty pace intervals between men. My intention was to deliver a short but heavy artillery concentration and then smash into the position five hundred yards northeast of Mount Cosna summit.

Fire adjustment proved to be a slow process, but I finally got the center of impact of an Austrian howitzer battery onto the Rumanian positions, only to hear that all the artillery had been ordered to cease fire for the remainder of the day because of changes in position and ammunition shortages. Meanwhile, the Rommel detachment reached the southeast part of the hollow despite lively Rumanian artillery fire, for the enemy did not miss the descent of seven hundred men. We found ourselves among clumps of bushes some three hundred yards from the enemy obstacles and out of his line of sight. One man was wounded slightly during the descent. I went down to the detachment and found that telephone wire had been laid.

The situation did not seem too promising and attacking an alerted enemy without adequate artillery support was out of the question, for the wired-in positions were much too strong. A daylight withdrawal over the steep northeast slope of Mount Cosna was equally unattractive in view of the excellent enemy observation and his ability to punish us severely with artillery and machine-gun fire. The men could run down the slope, but would move very slowly uphill and would offer excellent targets for the Rumanian artillery and machine guns. Heavy losses were unavoidable should the enemy decide to plaster our depression with artillery and trench mortar fire.

In spite of the unfavorable situation, I decided to attack the Rumanian positions without artillery support. I knew my men could do it and it was better to be a hammer than an anvil! Skilled scout squads reconnoitered the hostile obstructions and the positions behind them. In order to run in under the expected hostile artillery fire, I moved the detachment up through the bushes to within two hundred yards of the enemy position and made my preparations for the attack in the small draws in that area. The machine gun companies located some positions up on the slope to the right from which they could deliver supporting fires. The results of reconnaissance were not unfavorable and the enemy gave no signs of having noticed our offensive intentions. I was just about to order the two machine-gun companies to move into their reconnoitered positions, when the following order came over the telephone from Major Sprösser:

"The Russians have broken through in the Slanic valley to the north, and are now apparently about to come up on our rear. The Rommel detachment and the two Bavarian battalions withdraw immediately to the ridge half a mile west of Mount Cosna."

The group staff was heading there and I was ordered to transmit this order to the 1st and 3d Battalions of the 18th Bavarian Reserve Infantry Regiment, and to cover their retreat.

A bad situation!

In my opinion the most difficult maneuver was the daylight withdrawal from the hollow in full sight of the enemy. If the Russians spotted our movements he would be sure to hit us with machine gun and artillery fire and he would probably launch an attack that would cost us too many casualties. The Rumanians were less of a worry, for I hoped to reach the ridge ahead of them. My alternate plan, in case of failure, was a quick, fierce thrust to swing them off the ridge.

Under command of Lieutenant Werner of the Württemberg Mountain Battalion, I ordered the two Honved companies to climb the northeast slope of Mount Cosna, with the mission of reaching the summit; I took the four remaining companies and we worked our way through the brush, first in the direction of Hill 491, then toward Headquarters Knoll. Rumanian machine-gun fire hit us in the vicinity of Hill 491 and caused a few casualties.

Once in the vicinity of Hill 491, I ordered the 3d Company to occupy the lower part of the ridge running between Elevations 788 and 491 and instructed them to be prepared to receive the two Bavarian battalions which had received Major Sprösser's orders by officer messenger. Unfortunately, the wire communication with group headquarters had been interrupted. By chance, I overheard a conversation near Point 491 which indicated that recent developments led Major Sprösser to take a more favorable view of the situation.

Thereupon I moved the 2d Company by the shortest route to the ridge leading northward from Headquarters Knoll. The company was to organize the ridge six hundred yards north of Headquarters Knoll and was to provide security and reconnoiter in the direction of the Slanic valley. I ordered all units, except the 3d Company, to march to Headquarters Knoll while I stayed with the 3d Company. In the course of the next hour both Bavarian battalions succeeded in disengaging themselves from the enemy.

As soon as I saw they were being successful, I took the 3d Company and started for Mount Cosna. The 1st and 6th Companies were still on the Cosna summit, which the increased bombardment had turned into a field pockmarked with all sizes of craters. I left the 3d Company on the summit as a reinforcement to the garrison, reported to Headquarters Knoll, and asked permission to go to the hospital, for I was completely exhausted and did not feel able to continue in command. The bandage on my left arm had not been changed since morning, and so I gave up command of my companies and went to get some rest near headquarters. It was a pitch-dark, warm summer's night.

Chapter 9

FURTHER OPERATIONS AT MOUNT COSNA

I: The Defense, August 14-18, 1917

Shortly before midnight, Major Sprösser summoned me to headquarters where I found a large number of officers. Major Sprösser told me that the situation was most unfavorable. Reports from isolated units of the Hungarian 70th Honved Division (3d Troop Imperial and Royal Uhlans, 1st Troop, Imperial and Royal Dragoons, and 1st Honved Company) informed us that during the afternoon strong Russian and Rumanian forces had broken through the division in and to the north of the Slanic valley and were preparing to move south against the Mount Cosna-Ungureana ridge. We had to reckon with the assumption that, under certain conditions, the Sprösser Group would be cut off, for we had no troops to our rear short of the Ungureana. I was asked to express my views.

My opinion was that a night attack against the Mount Cosna-Ungureana line was most unlikely, and that the earliest attack would come at dawn which was only four hours away. With the group's five battalions I considered it possible to hold the Mount Cosna-Ungureana line against all comers, for the retention of this position was vital to the general situation. Under no circumstances would I surrender supinely the territory taken with so much resourcefulness, skill and blood, simply because of alarming reports.

I proposed the following regrouping be effected without delay:

"The mountain battalion assumes the defense of Mount Cosna, Headquarters Knoll, and the ridge as far as Hill 674. The other battalions of the group seize and hold the ridge between 674 and Ungureana. All units push through reconnaissance and security elements toward the Slanic valley."

For the disposal of the mountain battalion I proposed:

"Combat outposts, a rifle platoon reinforced by machine guns, occupy the south portion of Mount Cosna. The crater-field on the summit is not occupied. Reconnaissance to the southeast and east. A platoon and a heavy machine-gun platoon occupy the Headquarters Knoll and prevent the enemy from occupying Mount Cosna summit. A rifle company occupies each of the two ridges descending to the north between Mount Cosna and Elevation 674. Reconnaissance and security to the north. All remaining companies are assembled just southwest of Headquarters Knoll and held at the commander's disposal."

Major Sprösser accepted my recommendations and urged me, since

I took the terrain by attack, to defend the Württemberg Mountain Battalion sector. The seriousness of the situation, concern for my mountain soldiers, and last but not least the stimulation of the difficult task, led me to shoulder this new burden.

Oral group order initiated the regrouping which was executed without further delay. I had the following for the defense of Mount Cosna sector: 1st, 2d, 3d, 5th, and 6th Rifle Companies, and the 3d Machine-Gun Company of the Württemberg Mountain Battalion, and the 3d Company of the 11th Infantry Reserve Regiment with six heavy machine guns.

The group staff retired to the oak woods a mile northeast of Ungureana (the salient). My company commanders and I discussed the situation in detail and I issued the following order:

"The 3d Company moves immediately from Mount Cosna to Headquarters Knoll and sends a platoon without packs, but reinforced by six light machine guns of the 3d Company of the 11th Infantry Reserve Regiment, to relieve the 1st Company on Mount Cosna. This platoon (reinforced) occupies the wooded southern ridge and reconnoiters toward the hostile position east of Mount Cosna. In case of attack, the platoon holds its position as long as possible and retires on Headquarters Knoll only if threatened by encirclement. I shall give oral instructions to the platoon commander at a later time.

"Another platoon of the 3d Company, as well as Albrecht's heavy machine-gun platoon, dig in on Headquarters Knoll so as to cover the crater-field on Mount Cosna and the west slope with fire. They will prevent the enemy from crossing the bare part of Mount Cosna by day and threaten the combat outposts on the left flank.

"The 2d Company occupies the small knoll seven hundred yards north of the Headquarters Knoll (later called the Russian Knoll), reconnoiters toward the Slanic valley, maintains night contact with the combat outposts on Mount Cosna by scout squads. The company will build large campfires on Mount Cosna's northwest slope in order to deceive the enemy and divert his artillery fire. These fires will be kept going all night.

"The 5th Company, reinforced by a heavy machine-gun platoon occupies the knoll half a mile northeast of Hill 674, and prepares for all-around defense. It will reconnoiter toward the Slanic valley and maintain contact with the 2d Company and the neighboring troops in region of Hill 674, or Piciorul. To deceive the enemy and divert his artillery fire, the company builds large fires in the hollow half a mile northwest of the Headquarters Knoll and keeps them burning all night.

"A platoon of the 3d Company, Aldinger's machine-gun platoon, 1st and 6th Companies of the Württemberg Mountain Battalion, and the 3d Company of the 11th Infantry Reserve Regiment go into reserve areas between Headquarters Knoll and the descending slope a quarter of a mile to the southwest. Security and reconnaissance in the direction of Grozesti. More detailed orders will be issued later.

"Detachment combat outpost sixty yards west of Headquarters Knoll. Communications platoon lays wire connection to the combat outposts and to the 2d and 5th Companies."

While the leaders repeated their orders much activity began. The Bavarians and Honveds moved back followed by the companies of the Württemberg Mountain Battalion. Sleeping was out of question, for individual orders had to be issued on the spot to meet particular situations. It took three hours to get the companies in their new positions. The campfires on Mount Cosna and in the hollow northwest of Headquarters Knoll were burning, and contact with the various units had been established. The reserve units rested while those in position dug themselves in. The reconnaissance detachments made no alarming reports.

My staff consisted of Lieutenant Schuster as adjutant and Lieutenant Werner as administrative officer. Toward 1700 some artillery observers (including the Hungarian First Lieutenant Zeidler) arrived, and I went with them to the combat outpost on Mount Cosna. We reached Allgauer's platoon (3d Company) just as the sun was rising above the horizon. In accordance with orders, Allgauer had located his platoon on the sharp ridge leading south from Mount Cosna summit. The position was so organized as to have its flank on the edge of the thick forest some two hundred yards south of Hill 788 (2500). The Rumanian positions were visible through the haze, and were on a bare ridge some 350 feet between and about a half mile away. We saw the sun's rays reflected from the helmets of the large garrison, but there was no firing and the men who had had no rest, were asleep in their freshly dug foxholes, leaving only the sentries to keep sharp watch in the enemy's direction. The slope in front of the platoon position fell sharply to the east and was covered with short shrubbery. The ridge itself as well as its western slope, was

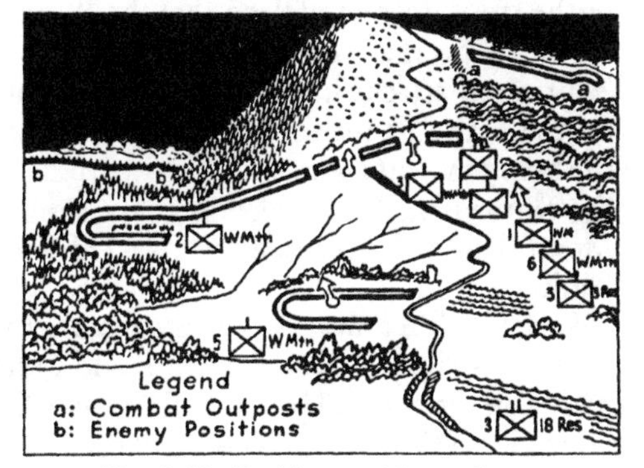

Sketch 29: Positions at Mount Cosna, August 13, 1917.

covered with large trees and had little or no protective undergrowth.

While discussing emergency barrages and harassing fire with the artillery observers, the various sentries reported: "The Rumanians are leaving their positions in a skirmish line and are advancing toward Mount Cosna." Shortly thereafter violent Rumanian machine-gun fire was directed at the Mount Cosna ridge line, and heavy artillery began to fire on Headquarters Knoll. I got through to our artillery and requested harassing fire on the Rumanian positions east of Mount Cosna, from which increasing numbers of troops were coming. In the meantime this report arrived: "Strong enemy has been located just in front of the line of combat outposts and is climbing the ridge from the right." The bursting of numerous hand grenades, lively carbine and machine-gun fire confirmed this report. Retribution was being exacted for inadequate security measures on the steep eastern slope. By telephone I ordered the reserve platoon of the 3d Company and Aldinger's machine-gun platoon forward at the double to reinforce the combat outposts. This order was followed by a request to the group for emergency barrages. I made a tour of the front lines and found that the Rumanians had secured a foothold on the ridge and were delivering flanking fire on our combat outposts. All frontal attacks had been beaten off and our artillery was smashing up the many Rumanian reinforcements on the bare slope. On the left the heavy machine-gun and rifle fire from Headquarters Knoll prevented the Rumanians from crossing either summit or the northwest slope of Mount Cosna. This fire also protected our combat outposts on the left flank.

I ordered Technical Sergeant Allgauer to hold the position at all costs until the arrival of reinforcements, and I ran back to speed the reinforcements on their way. Heavy shells were still hitting Headquarters Knoll, where I met the two platoons preparing to get under way and we hurried forward at the double. The noise of battle had increased considerably and we hoped that Allgauer was holding on!

In the saddle between Headquarters Knoll and Mount Cosna we encountered several light machine-gun crews of the 3d Company, 11th Infantry Reserve Regiment, who were part of Allgauer's platoon. Apparently it had become too hot for them, up front, and I treated them with scant consideration and took them along with me.

A hundred yards east of the saddle we saw Allgauer's entire platoon coming toward us; Allgauer reported that large Rumanian bodies had pushed up the slope and that these and the strong fire from below on the right had compelled him to give up the position.

I was in no frame of mind to surrender Mount Cosna at so cheap a price, and organized my forces for a counter thrust. Lieutenant Aldinger took two heavy machine guns and went into position in the woods on the right and kept the ridge, hitherto occupied by Allgauer's platoon, under steady fire. Simultaneously we climbed the ridge and passed through dense bushes before reaching the ridge line. Having arrived there, we charged ahead and swept the surprised enemy from the ridge and drove him down to the east; we also seized the promontory down on the right.

But the Rumanians were tenacious and did not let go. We clearly heard the commands of the enemy leaders down below us on the arched slope, and soon bitter hand-grenade battles began at various places. The slope was so steep that our hand grenades did not burst among the Rumanians lying in readiness 125 yards below us, but actually fell even farther before detonating. To reach him with a carbine meant exposure of head and shoulders, a procedure most disadvantageous at our short ranges. Losses began to increase and Dr. Lenz had much work to do in the front line.

The mountain troopers fought with exemplary bravery; many wounded returned to the firing line after having their wounds bandaged. All Rumanian footholds on the ridge were immediately wiped out by counterattacks, mounted by the nearest group of mountain soldiers. The hard battle, replete with casualties, lasted for several hours and ammunition and hand grenades gradually became scarce while hostile artillery fire against Headquarters Knoll increased. Telephone connection between Headquarters Knoll and the combat outpost position was shot away. If I wanted to hold on to my combat outpost positions, then the time had come to reinforce them with additional forces, ammunition and hand grenades. In order to expedite matters (telephone communication was lacking), I put Lieutenant Stellrecht, the 3d Company Commander, in command and ordered him to hold at all costs while I hurried back to Headquarters Knoll, where I found the following situa-

Sketch 30: Defense of Mount Cosna. August 13, 1917.

tion: The platoon of the 3d Company and Allbrecht's heavy machine-gun platoon had used up nearly all their ammunition against the enemy, who was threatening the left flank of the combat outposts from the crater-field on Mount Cosna. My reserve companies (1st and 6th Companies of the Württemberg Mountain Battalion, as well as the 3d Company 11th Infantry Reserve Regiment), had occupied the south slope of Headquarters Knoll on their own initiative because strong enemy forces were reported ascending toward Headquarters Knoll through the ravines from Grozesti.

Before I had units of these companies ready for use, we had reports that strong Rumanian forces were advancing both from the south and from the north against the saddle between Headquarters Knoll and Mount Cosna, and that the combat outposts had abandoned Mount Cosna and were retiring on Headquarters Knoll. In the next few minutes (I still had no men at my disposal), the noise of battle approached dangerously near to Headquarters Knoll and the riflemen of the 3d Company were retiring on the Knoll hard pressed by a superior and aggressive enemy. They brought their dead and wounded (including Lieutenant Hummel) back with them, for they had no intention of allowing anyone living or dead to fall into enemy hands. Hand grenades and machine-gun ammunition had given out, carbine ammunition had become short and they were threatened with encirclement from either flank.

The lack of ammunition and hand grenades made it most difficult to stop the attack of the Rumanian masses against Headquarters Knoll. The heavy machine gunners had to defend their positions with pistols and hand grenades, and the few runners of my staff were used at threatened places. Violent fighting raged along the entire front. At that moment I discovered large numbers of Rumanians in the wooded part of the depression, seven hundred yards northwest of Headquarters Knoll. I informed the 2d and 5th Companies by telephone regarding the new danger which threatened their flanks and rear.

In all parts of the sector violent fighting was going on and a withdrawal was out of the question. What would happen on Headquarters Knoll when the ammunition was completely expended? With the dominant position in enemy hands, the entire battalion would be in a most precarious predicament and our entire defense would collapse. We could not allow that to happen. Telephone connection with the group still existed and I described our current crisis and urgently requested immediate reinforcements including small arms and ammunition. I stressed the fact that time was not to be wasted. The worries of the next half hour were indescribable, but at our eleventh hour the 11th and 12th

Companies, 18th Bavarian Infantry Reserve Regiment, and a heavy machine-gun platoon came to our assistance. The 12th Company with the heavy machine-gun platoon, went into position on Headquarters Knoll and I kept the 11th Company in reserve on the slope three hundred yards west of Headquarters Knoll, where I also located the detachment command post. From there I had an excellent view of the whole battle-ground.

I used the reserve company to resupply the forward positions with ammunition and hand grenades. All troops not actually firing at the enemy plied their spades with great vigor. The machine-gun fire from dominating positions on Mount Cosna was most annoying to those on Headquarters Knoll and on the ridge. I withdrew Aldinger's heavy machine-gun platoon from the front line and put it in a defense zone in the neighborhood of the detachment command post. Furthermore, I established ammunition supply points and put my supply system in order.

The battle for the Headquarters Knoll and the Russian Knoll continued for hours without pause. The enemy repeatedly hurled new forces against our thin lines, and Rumanian artillery concentrations on the slope just west of Headquarters Knoll prevented contact with the front line and tore up our telephone connections. But the Bavarians and Württembergers up front held their positions, and our own artillery did a good job during the course of the day in giving us emergency barrage at all threatened points. Its shells thinned the ranks of the Rumanians who were lying in dense masses on their lines of departure.

The fire of several batteries was made available for use in destroying the enemy, who was moving back in a depression half a mile northwest of Headquarters Knoll. Heavy concentrations were prepared and were available on call. In spite of the excellent artillery cooperation, I still lacked observers up forward and was also in need of wire communication with the artillery command posts.

By noon there were mountains of dead and wounded Rumanians in front of Headquarters Knoll, but the 12th Company of the 18th Infantry had also suffered heavily and had to be refilled by units of the 11th Company. Later still more units of the 11th Company had to be used to fill gaps in the 2d Mountain Company.

The defense arrangements on Headquarters and Russian Knolls included light front-line garrisons, with strong counterattack groups assembled under cover in vicinity of the more threatened portions of the position, with the mission of ejecting the enemy from any point where he managed to effect a breakthrough. This type of defense lent itself to our particular terrain.

In the afternoon, the 10th Company of the 18th Infantry arrived as additional support and I ordered it to dig a communication trench from Headquarters Knoll to the detachment command post. The Rumanians switched their main attack against Russian Knoll. There Hügel's platoon had organized itself for all-around defense in some old Rumanian positions, and was hit hard from the north and east by an enemy who outnumbered him ten to one. The enemy tried repeatedly to regain positions whose installation had cost him weeks of work. The Aldinger's heavy machine-gun platoon at the detachment command post spoiled all enemy attacks from the west against Hügel's platoon, and the 2d Company gallantly held its ground.

The battle raged in undiminished fury and almost without interruption into the late afternoon. For the third time I ordered ammunition and hand grenades replenished in the front line. Through the smoke clouds of our heavy shells (305mm guns were used in defensive fires), we saw more and more fresh Rumanian troops descending the slopes of Mount Cosna in our direction. When the 2d Company reported that it had dwindled to such an extent that it was obliged to retire from Russian Knoll, I sent the remaining units of the 11th Company of the 18th Infantry to its support. At the same time I ordered two heavy machine-gun platoons to prepare destruction fire on Russian Knoll. When these preparations were completed, I ordered the 2d Company to vacate Russian Knoll rapidly. As expected, the hostile forces stormed up on the bare knoll in a dense mass; at the same moment the destruction fire of the heavy machine-gun platoons struck among them and mowed them down like ripe corn. In full flight, the survivors fled the dangerous knoll and shortly thereafter the reinforced 2d Company was again in possession and was allowed a brief respite.

Somewhat later the Rumanian forces, which we had observed for hours in the depression half a mile northwest of Headquarters Knoll, started moving up the slope to the south. The previously prepared artillery fire was requested and had excellent effect; it drove the enemy back into the lower woods. Thus the rifle and machine-gun fire prepared for the reception of this enemy by the 2d, 12th, and 5th Companies and the three heavy machine-gun platoons was unnecessary.

During the battle, message after message came from the front line. The adjutant and the administrative officer had all they could do in executing hasty requests for protective fire, supplying ammunition and keeping Major Sprösser informed as to latest developments. Double wire lines were laid to the more threatened points and to Major Sprösser's command post and kept in repair by the untiring communications men, a most dangerous job in view of the almost continuous machine-gun and artillery fire which kept searching the area.

In spite of the heaviest losses, the Rumanians continued their attacks into the night, but failed to gain a foot of ground. When the noise of battle died down in the night, we heard the groans and laments of the wounded all along the front. Our stretcher-bearers were fired on as they attempted to help some of the unfortunate men, and had to return without accomplishing their mission.

In my opinion, the enemy would repeat his attacks on August 14 with still stronger use of artillery and fresh infantry forces. Such serious losses as we suffered on August 13 could not be repeated. Therefore, I ordered the short hours of the night employed in fortifying our positions and reorganized the defense at various places. With the company and platoon leaders, some of whom still had little combat experience, I laid the trace of the main line of resistance on the ground and prescribed the type of construction to be used in the defense installations. During the night fields of fire had to be cleared at various points; furthermore, in the arrangement of the rifle and heavy machine-gun nests, it must be remembered that the enemy was able to cover them from dominant positions on Mount Cosna. The 234th Pioneer Company (Engineers) which was brought up and assigned to me just before dark, was given the extensive work on Headquarters Knoll.

Only by midnight were all portions of the extended sector assigned to units which began work immediately. I was exhausted when I reached my command post, but a warm lunch refreshed me. Sleep was out of the question. The wounded had to be attended to, ammunition and hand grenades had to be supplied to the companies in the front line and to the depots before daybreak; provisions had to be brought up to the individual companies; the communications platoon had to lay a double line to the artillery fire direction center, and then the combat report for August 13 had to be forwarded to the Sprösser group.

We finally finished all this work and at 0400 I tried to get some sleep, but it was so cold that I gave up the idea; so I took Lieutenant Werner and inspected the night's work in the early dawn light. I had not had a chance to remove my shoes for more than five days and, as a result, my feet were badly swollen; also, I had had no opportunity to renew the bandage on my left arm or to change the blood-stained overcoat hung around my shoulders and my likewise blood-stained trousers. I felt very debilitated, but the weight of responsibility was such that I did not consider going back to the hospital.

At daybreak on August 14 a Honved infantry company with light machine guns arrived and I ordered it to relieve the 1st and 3d Companies; I put these two companies in reserve just west of my command post. The 11th and 12th Companies of the 18th Infantry had taken over respectively the Headquarters Knoll position and the position astride the ridge road. I left the 10th Company of the 18th Infantry in its position in the woods three hundred yards west of Russian Knoll. It had pushed its security elements to the north and northwest in the direction of the Slanic valley. We were ready and felt the battle could recommence.

During the whole forenoon the Rumanian artillery bombarded our positions on Headquarters Knoll, the ridge road and on Russian Knoll very actively, but caused little damage. In all sectors work was carried on busily and the positions were further improved, so that a strong Rumanian attack on the whole front at noon was easily repulsed.

The 2d Company on the Russian Knoll suffered heavily from the fire of a Rumanian battery, located in an open position about a mile away. We had no artillery observers in our sector and our attempts to adjust fire by telephone were unsuccessful, and we did not succeed in neutralizing this particular battery. The enemy strengthened his positions on the west slope of Mount Cosna, and the hostile wounded continued to groan and moan in front of our lines. Our own losses on August 14 were slight and August 15 was also a quiet day. I took advantage of these two days to have two draftsmen reproduce and grid a sketch and map of the Mount Cosna terrain, which I had drawn to the scale of 1:5,000. The group artillery commander and the artillery observers received copies; the artillery made sufficient copies so that distribution was made to include all batteries. A grid map or sketch greatly facilitates adjustment of fire in mountainous or wooded terrain, where it is often difficult to select visible aiming points or targets by map study alone. For example, I notified the artillery: "Request emergency barrage in squares 65 and 66." If now the requested fire was outside them, then it sufficed to say: "Emergency barrage requested in squares 65 and 66 is in squares 74 and 75," in order to bring the fire quickly into the desired region. Combat information within one's own unit and group was considerably simplified. For example: "Rumanian battery located in square 234a."

In the night of August 15 the mortar company under Lieutenant Wöhler arrived, made a night reconnaissance and began emplacing its mortars. Captain Gössler came forward to spell me, for I had not had any rest for a week. Command remained in my hands. In the afternoon the 4th Company arrived as additional reinforcements, and I found myself in command of sixteen and a half companies, or with more strength than an entire regiment.

The 11th Infantry Reserve Regiment was on our right, but our left was up in the air. Brigade was trying hard to establish a continuous front, but insufficient troops were available for the task. The defense of the steep, wooded slopes of the Slanic valley required an enormous force.

Following a period of oppressive heat, a heavy thunderstorm broke on August 16 and the thunder echoed and reechoed in the mountains, accompanied by pouring rain from the low-hanging clouds. The covered Rumanian positions west of the command post gave shelter to the staff and the detachment reserves, but not for long, for they soon filled and had to be vacated. With lightning flashing all around we lay in the open, sopping wet, when a sudden hail of artillery of all calibers drowned out the noise of thunder. Violent rifle and machine-gun fire began up front accompanied by hand-grenade bursts. The Rumanians hoped to surprise us in the storm! I began to wonder if the front still held or had been overrun. The rain beat into our faces so sharply that visibility was down to a few yards; I decided to act rather than await reports.

Headquarters Knoll was the focal point, and in a few minutes I reached a point just to the west of the Knoll. With me was the 6th Company, bayonets fixed and ready to counterattack. Our emergency barrage plowed up the region in which the Rumanian masses were attacking; a combat telephone line connected us with my staff and thus with all points in the sector. The Rumanian attack collapsed everywhere and night put an end to the confusion of battle in the streaming rain. Only after suffering heavy losses in dead and wounded did the enemy retire from the area in front of our positions.

Upon return to my command post at the conclusion of fighting, I found the place on which we had pitched our tents plowed up by heavy shells. Under these conditions I moved the command post three hundred yards to the right. We dried our wet clothing on our bodies by the heat of a fire tended by Rumanian prisoners. We were in fine spirits!

Observations: The task of the Württemberg Mountain Battalion on August 13 to defend parts of Mount Cosna and the high ground immediately to the west, was exceptionally difficult. With no contact on either flank, the battalion had to prepare for strong hostile attacks not only on the front but also on both flanks. Then, too, the very irregular, thickly wooded terrain on both sides of the bare ridge favored hostile approach to within attacking distance. Furthermore, the Rumanian artillery was in a position in a semicircle around the Württemberg Mountain Battalion.

Under these circumstances a defense in great depth and a retention of strong reserves was desirable.

Active combat reconnaissance toward the south, east and north was necessary even before daylight, in order to determine the hostile offensive intentions. Further, the unsurveyable terrain out in front of our own positions had to be kept under constant and sharp observation.

Where that was not done, as at the combat outposts, unwelcome surprises were experienced.

The fighting at the combat outposts was very difficult. To be sure, they had a field of fire from the sharp ridge of Mount Cosna far into the open hostile territory, but the arched, steep and densely covered slope in the immediate foreground could not be covered with fire. Their security measures were inadequate. It was here that the Rumanians made preparations for a daylight attack with strong force. Their attack was a complete surprise for the combat outposts.

Machine-gun and rifle fire from Headquarters Knoll against the bare summit and lightly wooded west slope of Mount Cosna managed to protect the left flank of the combat outposts for a considerable time, and it was only when ammunition gave out on Headquarters Knoll that the enemy managed to set foot on Mount Cosna.

Under the quickly organized fire support of a heavy machine-gun platoon, it was possible to regain the last line of the combat outposts without suffering much in the way of casualties. The fire and movement of the assault squads were in complete unison here.

The fighting along the outpost line and for Headquarters Knoll are excellent examples of the rapidity with which ammunition becomes exhausted at the focal points of combat. In such cases (especially in the mountains), resupply must be established at the earliest possible moment. Besides that, a reserve of ammunition and close-combat weapons must be on hand in the battalion. The battalion supply point must be constantly informed as to the amounts of ammunition on hand in the forward line and must get resupply started. Supply worked well in the course of the fighting on August 13.

Reserves were urgently needed during the heavy fighting on August 13; without them the position could not have been held; again and again, losses in the principal combat zone had to be replaced by reserves. The supply of ammunition and close-combat weapons was brought to the front line by the reserves. During the battle, a communication trench had to be dug by a reserve company from the battalion command post to Headquarters Knoll, the focal point of the fighting. Without the trench, supply would have been brought up only with heavy losses in face of hostile fire from the dominant Mount Cosna position.

Even at the beginning of the defensive battle the Württemberg Mountain Battalion was deeply echeloned in the main area of combat. The 5th and 2d Companies and the forces disposed on Headquarters Knoll could support each other with fire. During the battle, reserves at the foci of the fighting (Headquarters Knoll and Russian Knoll), deepened the defense area. It would have been a mistake to put everything in the front line of nests; the losses were heaviest there, and they would have been still greater if the garrison had been stronger. It is easy to break a line.

The cooperation with the artillery was very satisfactory on August 8. Of course, an artillery liaison agency or advanced observers in the battalion sector would have accomplished still more advantageous results. The grid sketch prepared during the defense was very valuable; it corresponded to the present protractor and scale or plotting board.

II: Second Attack on Mount Cosna, August 19, 1917

After several days of heavy fighting, our neighbor on the left (70th Honved Division), succeeded in advancing north of the Slanic valley, and a continuation of the attack on a broad front on both sides of the Ojtoz and Slanic valleys was planned for August 18. Mount Cosna was to be attacked once more and seizure of positions to the east was part of the general plan. Then the command hoped to effect the breakthrough. For the attack against Mount Cosna we had Madlung's Group (22d Infantry Reserve Regiment), on the right and Sprösser's group (Württemberg Mountain Battalion and 1st Battalion, 18th Infantry), on the left. On August 17 I was ordered to complete all attack preparation for the front line units of Sprösser's Group; I was also instructed to acquaint the regimental and battalion commanders of Madlung's Group with the terrain over which they were to attack. As a result, I was on the go from dawn till dark.

When I returned to my command post I learned that, following a strong artillery preparation, the Rumanians had launched an attack against the Piciorul from the Slanic valley, that is, from the left-rear of our positions. They were opposed by elements of the 18th Bavarian Infantry Reserve Regiment, and the sounds of battle indicated that the Rumanians were making considerable progress. My flank and rear appeared threatened, and I was afraid of being cut off from the group. As a preventative measure I hurried part of my reserves (two rifle companies, one machine-gun company), at the double to the vicinity of Elevation 674 and concealed them there in clumps of bushes, ready for a counterattack. Telephone communication was established with my command post, and Group Headquarters reported that the Bavarians on Piciorul had stopped the attackers; consequently, my reserves were not committed.

The attack against the Mount Cosna was postponed one day. During the night of August 17-18, the companies in the right-hand part of the sector were relieved and moved to the second line. On August 18 the 2d Company, with units of the 18th Infantry, cleared the Rumanians from the ridge six hundred yards north of Russian Knoll. On this rainy day

I roamed the territory around Russian Knoll with German and Austrian artillery observers and perfected plans for artillery support for the attack on August 19 against the northern part of Mount Cosna.

Before daybreak on August 19 the assault troops of Sprösser's Group assembled in the draw northwest of Headquarters Knoll. A new grouping had been organized. I led the assault companies consisting of the 1st, 4th, and 5th Companies, 2d and 3d Machine-gun Companies, an army assault detachment, and an engineer platoon. Captain Gössler was to follow in the second line with the 2d and 6th Companies and the 1st Machine-gun Company. Sprösser's Group also had the 1st Battalion, 18th Infantry at its disposal.

My detachment assembled in the clumps of bushes and strips of wood just west of the Russian Knoll, while the other units of Sprösser's Group assembled farther to the west. The enemy had built a continuous trench system and had erected obstacles in front of it on the ridge running from Mount Cosna summit northwest in the direction of Elevation 491. By sharp observation with field glasses, we could see parts of the position and the obstacles between the bushes.

According to division orders, this position was to be taken after a one-hour artillery bombardment. After another hour's bombardment, the particularly strongly fortified position half a mile east of Mount Cosna summit was also to be taken. This was the position we had come up against on August 13. I intended to break into the hostile Mount Cosna position during the artillery bombardment, advance through this position a little, and then shift our artillery fire on the second Rumanian position and start attacking it.

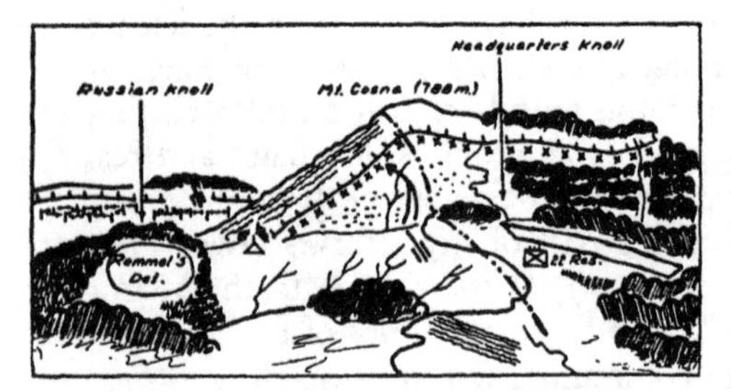

Sketch 31: Situation on August 19, 1917.
View from the west.

The summer weather on August 19 was magnificent. There was no fighting in the Mount Cosna sector during the early morning hours; the assault troops were hidden in the bushes. Toward 0600 I sent out Technical Sergeant Friedel (5th Company) with ten men and a telephone squad, explained my plan of attack and gave them the following mission:

"The Friedel scout squad, under cover of bushes and depressions, climbs from Russian Knoll through the ravine to the east into that hollow over there [indicated on the terrain] toward the site of the planned breakthrough, and reconnoiters the obstacles in front of the position.

Wire cutters are to be taken along, and continuous contact even during the assault is to be maintained with the detachment command post through the telephone squad."

A half hour later I saw Friedel's scout squad climbing the west slope of Mount Cosna. In the meantime I had located Rumanian sentries in trenches near the breakthrough site. The telephone connection with the Friedel scout squad was in order and I could keep him informed of all new developments in the hostile position above him; I could also tell him at any time how far he was from the hostile position, and could guide him to the intended breakthrough site. It did not take him long to reach the hostile obstruction.

When the Rumanian sentries in the trench became uneasy—apparently they had seen or heard the scout squad—I withdrew it two hundred yards from the wire entanglement and had Lieutenant Wöhler's mortar company open fire on the breakthrough point from positions in our rear. Shells were soon bursting around the hostile sentries and they either dove for cover or moved laterally out of the danger area. While Wöhler's company was using fire for effect, I ordered Friedel to cut a passage through the hostile obstacles fifty yards from our shell bursts. This job was accomplished rapidly and without disturbance.

The artillery preparation was scheduled for 1100, and at 0900 we started out with the detachment on the path taken by Friedel and marked by the telephone line. The slope from Russian Knoll down to the ravine on the east was in the sun, and the bushes gave insufficient cover and the Rumanians soon discovered the movement. In spite of increased intervals between men and a more rapid pace, Rumanian machine-gun fire caused a few casualties. On the other hand, Mount Cosna's arched west slope was defiladed from enemy fire and was not being observed by the enemy.

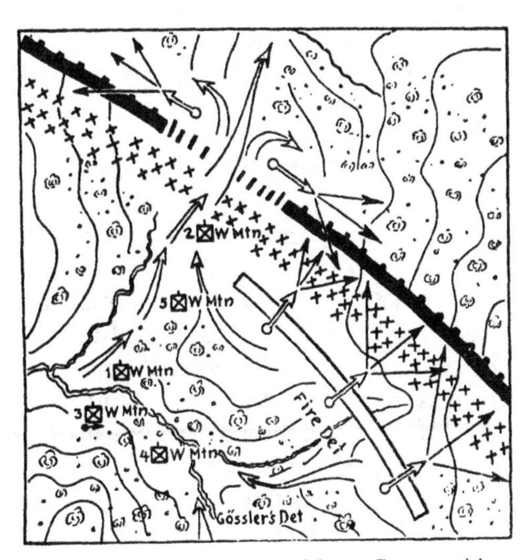

Sketch 32: Penetration of the Mount Cosna position.

When I reached Friedel with the head of the column, the hostile obstacles had been cut through up to the last few wires. During the advance of the detachment, Lieutenant Wöhler, who remained behind in observation on Russian Knoll, had kept me constantly informed of all developments in the hostile position. From time to time, upon my request, he had a few mortar shells fired for harassing effect.

I had the companies close up fifty yards away from our breakthrough point, and began to examine the possibilities for launching our attack from a line nearer to our selected breakthrough site. Gössler's detachment was moving up through the draw on our right. It was 1030 and the 1st Battalion, 18th Infantry was still climbing. My plans were to attack shortly after the artillery preparation began and this meant that I had to speed up my attack preparations.

The entire 2d Machine-gun Company and a platoon of the 5th Company were to deceive, divert and pin down the hostile garrison of the positions above those we were attacking. These units were to crawl into position under cover and fire only on order. Their left flank was just above the gap in the wire entanglement. A few seconds after this formation opened fire, Friedel's assault squad was to attack down the path through the wire, break into the position and block off both shoulders of the penetration. I was to follow on Friedel's heels with the rest of the 5th Company, Lieutenant Leuze's heavy machine-gun platoon, and the remaining units of my detachment. Following a successful penetration I intended to take the 5th Company and drive straight ahead, paying no attention to developments on either flank, and seize the ridge lying to the northeast. I was to be followed by the 3d Machine-gun Company, the 1st and 4th Companies, the army assault detachment and the engineer platoon.

Leuze's heavy machine-gun platoon was given the job of sweeping the hostile position from the place of breakthrough to the right (uphill) and left (downhill) with heavy machine-gun fire. All other units remained in reserve. The units used for deception were to follow us into the captured position as soon as possible; Captain Gössler and I agreed that his forces would follow behind me. Elements of the 1st Battalion, 18th Infantry, had the job of rolling up the enemy flanks on Mount Cosna from our breakthrough point in the direction of Elevation 491; the remainder of the battalion remained in group reserve.

Our artillery began to plaster the Mount Cosna positions before we had completed our attack preparations and before the other detachments had occupied their positions from which they were to exploit our breakthrough the 210mm and 305mm shells threw earthen geysers into the air, and earth and bushes spattered down. The mountain soldiers' hearts rejoiced at this powerful aid by the sister arm.

As prearranged, the breakthrough place itself, square 14, was free of our own artillery fire. Here our mortars did excellent preparatory work, and five minutes after the start of the artillery fire I gave my detachment the signal to attack.

The fire unit up above blazed away, followed a few seconds later by Friedel's assault squad running through the entanglement path and into the hostile position. Then the forward units of my detachment began to move. The sharp crack of hand grenades in our immediate vicinity drowned out the noise of firing up on the right. A few strides through smoke and haze and we were in the hostile trench. Friedel's assault squad had done a magnificent job, but unfortunately the valiant Technical Sergeant had been killed at the head of his men by the pistol shot of a Rumanian cavalry captain; yet these mountain soldiers pressed the attack with increasing violence and overwhelmed the trench garrison in close combat. The captain and ten men were captured, and then the assault squad divided to right and left to block off the shoulders. I reached the trench at the head of my detachment. Up on the right, the trench garrison was still resisting the attack supposedly coming from that direction. The configuration of the terrain and the dense undergrowth prevented these people from seeing that we had already smashed into their position; they did not see company after company moving at the double into the gap in their defensive system.

Confusion reigned; hand grenades burst all about, machine-gun and rifle fire criss-crossed the bushes and heavy shells burst in the immediate vicinity. The assault squad had cut a hole forty odd yards wide in the enemy position and had blocked off the shoulders. To smash up the enemy position downslope would have been easy, but I adhered to my original plan and left this to the following units. According to the original mission, the 5th Company was already pushing through the bushes in a northeasterly direction toward the nearest ridge. Shortly afterward Lieutenant Leuze opened fire with his heavy machine guns from the blocking-off positions on the enemy garrisons up and down slope, and I could push my way with the 5th Company into the defense area without too much concern. The adjutant announced the success of the breakthrough to Group and requested shifting the heavy-caliber artillery fire to the positions east of Mount Cosna in Sprösser's Group area.

Further in the defensive zone, we overran Rumanian reserves and took over one hundred prisoners; the rest fled. During the pursuit a few 305mm shells hit in our immediate vicinity and blew huge craters in the loamy soil, which could easily accommodate entire companies. This artillery did us no harm but made us somewhat nervous. We moved on. When we reached the ridge a quarter of a mile northeast of our line of departure, we saw our next objective lying far below us and some seven hundred yards away. German shells were striking in the draw ahead through which several Rumanian companies were retreating in disorder.

I quickly ordered a heavy machine-gun platoon to open fire on the retreating enemy, and ordered the remainder of the detachment to go down into the draw and pursue the retreating enemy. By telephone—the line had been brought along during the advance—I requested strong artillery fire in squares 76, 75, 74, 73, 72, 62, 52, and 42. True to my original plan, I intended to storm the second Rumanian position after a short artillery bombardment. It turned out differently!

The brief arrangement and telephone conversation required only a few minutes, and the first German shells were striking down below in the draw. The Rumanians were hurrying back to their new position over a narrow path toward the bushes as the fire of several heavy machine guns was striking among them. The effect was devastating at this short range. I questioned the advisability of taking advantage of the enemy's panic and overwhelming the second position by rigid pursuit. To be sure, we would come under our own artillery fire, but we had just escaped the bursts of our own 30.5cm. shells without injury. Nothing worse awaited us up forward.

We rushed downhill as fast as our legs would take us. Howitzer shells were striking in the draw and our heavy machine-gun fire still covered the enemy, who was pressing back into his positions through narrow passages in the wire. I was soon close on the enemy's heels with the forward units of my detachment. In the heat of battle we did not worry about the German shells striking to the right, left, and rear. The enemy ahead of us was in headlong flight; he did not yet know how close we were to him, for no enemy shots hampered our advance. Many dead and wounded Rumanians lay all about us. Our heavy machine guns shifted their fire to the left, we hurried through the obstacles and were soon in the hostile position. After a short rifle and hand-grenade struggle, the garrison fled and I quickly disposed the companies as they arrived:

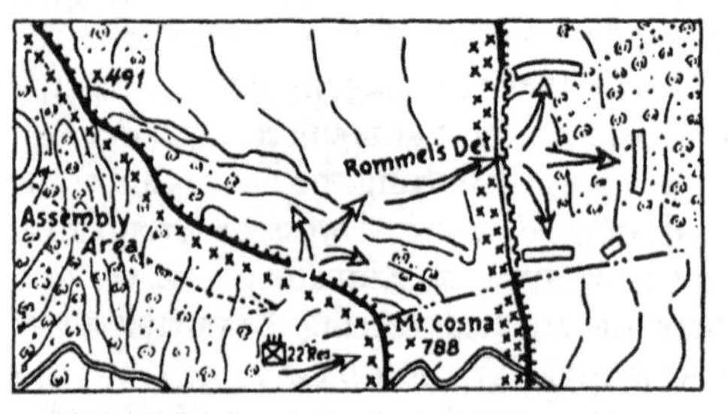

Sketch 33: Attack by Rommel's detachment on August 19, 1917.

"1st Company toward the east, 5th Company toward the north, 4th Company toward the south. Each company is to roll up the hostile positions for 160 yards, then halt and occupy and organize the position. Maintain active reconnaissance to your front."

After a few minutes I received word that all assigned objectives had been reached. On the right, in front of the 4th Company, the Rumanian garrison was most tenacious and even tried to regain its lost position by counter thrusts, but those were in vain for the mountain troops did not surrender that which they had once taken. To the east and north the Rumanians were retreating and even the artillery was evacuating its positions behind the ridge as rapidly as possible. The enemy was still holding out in the Madlung Group sector of Mount Cosna.

Over on the right, the enemy had occupied his secondary positions and, following the failure of his counterattacks, limited his efforts to holding these positions. A large gap in the hostile defensive system was visible to our left front. By throwing in all available reserves we could have broken through with relative ease.

Telephone communication with Group Headquarters had been established. My signal troops were marvelous and every bit as good as my assault troops. I quickly informed Group as to the situation up forward and requested the dispatch of all available reserves and adjustment of artillery fire on the second hostile position in the Sprösser sector. I learned that the hostile position on Mount Cosna to the right of Madlung's Group had not (1145) been taken. The immediate dispatch of Gössler's detachment and the 1st Battalion, 18th Infantry was promised.

I had to make most effective use of my available forces and could not overlook the possibility of counterattacks coming from Mount Cosna or from the south. The engineer platoon was given the job of improving the 4th Company's sector, and the latter extended its front to the east as far as a small wooded knoll from which a heavy machine-gun platoon began to fire on hostile batteries near Nicoresti (a distance of twenty-eight hundred yards), with the result that the batteries hitched up and evacuated their positions at the gallop. To the east, scout squads of the 1st Company pressed the retreating enemy downhill and through the light woods. To the north, the army assault detachment was rolling up the hostile positions beyond the line reached by the 5th Company. It moved forward rapidly. In the same direction and within two miles' striking distance was Tirgul Ocna. The town itself was under heavy artillery fire and we could see endless columns of vehicles halted near equally long trains in the station. We could have reached the town in thirty minutes, thereby cutting off the valley from which a large portion of the Rumanian forces drew its supplies.

Impatiently I awaited the arrival of Gössler's detachment and the 1st Battalion, 18th Infantry. According to information from the Group, both had been on the way for a long while. Minutes ticked by slowly, but nobody arrived and we could hear the sounds of combat to our right-rear where the possession of Mount Cosna was still being contested. Our captures had reached four hundred men and several dozen Rumanian machine guns. More than two hours had passed since the successful attack on the second position, and the Rumanians in the north were recovering from their fright and beginning to drive the assault detachment back. At the same time Rumanian artillery batteries in the region of Satul Nou fired some hundred rounds at the 4th Company, but most of them were "overs" and exploded harmlessly on the north slope of Mount Cosna. The enemy in the south did not counterattack, but his lively machine-gun fire forced us to hug the trenches and communication trenches. Sporadic hand-grenade fighting flared up, but the enemy failed to gain any advantage.

Gössler's detachment arrived at 1600 (four and a half hours after our initial attack), and his arrival coincided with a strong Rumanian counterattack from the north which forced us to commit the 6th Company in the gap between the 1st and 5th Companies. An attack against the valley was out of the question without adequate reserves. The hostile attack from the north was repulsed after a hand to hand fight.

At 1830 Group Headquarters reported that Madlung's Group had taken Mount Cosna (southern part) and was advancing east up the ravine to attack the secondary position.

Shortly before dark we observed the rearward movement of fairly strong Rumanian infantry formations near Nicoresti and Satul Nou. At the same time several successive trains pulled out of Tirgul Ocna and headed in an easterly direction. We made contact with the 22d Infantry Reserve Regiment whose left had taken the Rumanian positions at Hill 692 (2390). Hoping that we would break through to the plains on the following day, I put my detachment in an outpost line that extended well to the east and pushed my reconnaissance detachment as far as Nicoresti. In the north, a strong enemy still confronted the 6th and 5th Companies.

I was on my feet until midnight taking care of provisions for the troops, replenishing ammunition and preparing my combat report; then I lay down to sleep in a tent with Captain Gössler.

Observations: The attack on August 19, 1917, against the fortified wire-protected Rumanian positions lying half a mile apart, was a different kind of task for the Württemberg Mountain Battalion. Each position was to be taken after an hour's bombardment. The mountain soldiers broke through both positions with few losses while the artillery bombardment of the first position was going on and smashed into the second position along a seven hundred yard front and captured over five hundred Rumanians, thus paving the way for a breakthrough to the east, for it was improbable that the Rumanians had a third fortified and garrisoned position in the lowlands east of Mount Cosna.

Unfortunately, this great success could not be exploited because the reserves were too late in arriving and too weak in numbers.

The terrain called for unusual tactics. After penetrating the hostile positions close below Mount Cosna summit, we found it easy to break up the hostile positions on the steep descending northwest slope, particularly as this attack could be supported by heavy machine-gun fire from Russian Knoll.

It was vital that the leading elements penetrate a maximum distance in a minimum time and that there be no diversification of effort in smashing into the first positions. As a matter of fact, the forces were held together during the operations against the second line so that we might have a maximum of strength in hand for further exploitation when the reserves arrived.

The coordinated use of artillery, mortars, and heavy machine guns was the result of thorough prior planning. The mortar company nailed the enemy down at the breakthrough site even before the artillery preparation, and allowed Friedel's assault squad to cut a path through the wire. Artillery fire on the first position forced the enemy to cover during the breakthrough by the Rommel detachment while a machine-gun company and a platoon of the 5th Company fired on the enemy outside the breakthrough area and prevented him from interfering with operations.

The intense German preparatory fire on the first hostile position forced the strong Rumanian reserves to withdraw hurriedly to the second position. The Rommel detachment exploited the tactical situation and took up a vigorous pursuit and drove into the second hostile position, close behind the fleeing enemy, who was under heavy fire. In so doing the mountain soldiers ran the risk of coming under their own artillery fire, which could not be shifted so quickly.

III: Again on the Defensive

At 0300 on August 20 the enemy reopened the battle for Mount Cosna with violent artillery fire from numerous batteries. A large number of heavy shells struck in the vicinity of the command post and the reserve areas, and forced us to vacate the endangered areas and take shelter in the draw half a mile north of Hill 788. The enemy fire increased steadily with its bulk directed at the position captured by us east of Mount Cosna, where the Russians evidently expected us to be. I was very glad that I had a few of my men dug in there, for the fire soon converted this position into a mass of rubble.

At 0700 the enemy began to advance against the deep outpost occupied by the 1st Company and the draw near Nicoresti began to fill with Rumanians. The 6th Company, which was in the north, reported that it could observe attack preparations on the part of the enemy in its sector. All doubt vanished and we became convinced that the Rumanians intended to regain the territory lost the previous day. It was high time to shift to the defense.

A continuous line had to be formed in the rugged and wooded terrain, and the uncovered north flank required particular protection. I decided not to occupy the old Rumanian positions for they had been under heavy fire all morning, and the Rumanians had their range and knew their details most intimately. Defended, these positions would cost us too many casualties. In spite of the work involved and the short time remaining before meeting this strong enemy, I preferred to move the forward-slope position to the east and in the woods.

I issued the necessary orders on the spot and the companies dug in while the 1st Company outposts fought a delaying action. Digging was easy in the loamy soil, and the reserves helped the front-line units to dig their positions and their communication trenches so that all was in readiness when the combat outposts were finally driven into the positions. The initial assault was easily repulsed and the Rumanians began to dig in some fifty yards away from us. Rumanian artillery tried to reach our forward-slope positions but had to give up because of the danger of firing on its own troops. It therefore restricted itself to pounding the former Rumanian positions up on the ridge.

I had few worries regarding the eastern front (1st and 4th Companies), but the north and northwest sectors were a different story, for there we had a huge gap in our defense.

Our contact to the left (1st Battalion, 18th Bavarian Reserve Infantry Regiment) was along the northeast slope of Mount Cosna, on the ridge leading from Hill 491 (1620) to the summit, and the Rumanians took advantage of the draw and climbed up and reached the rear of our position. The 3d Company, hitherto in reserve, had to close the gap between the left flank of the 5th Company and the 1st Battalion of the 18th. It held in spite of a numerically superior enemy, poor defensive terrain, and miserable visibility. The battle increased hourly in violence, and during the day the enemy launched at least twenty assaults against us; some of them were preceded by a short artillery preparation and some were not. The Rumanian front was a semicircle about us and we had to rush our few reserves from one threatened point to another. The hostile artillery fire tore up the ridge but the mountain troops did not waver. Our losses were small in proportion to the enemy's for we had a total of twenty. I was so exhausted, probably because of the exciting activities of the past days, that I could give orders only from a lying position. In the afternoon, because of a high fever, I began to babble the silliest nonsense, and this convinced me that I was no longer capable of exercising command. In the evening I turned the command over to Captain Gössler and discussed the situation with him; after dark I walked down the ridge road across Mount Cosna, back to the group command post, a quarter of a mile southwest of Headquarters Knoll.

The Württemberg Mountain Battalion held its position against all Rumanian attacks until August 25th, when it was relieved by the 11th Reserve Infantry Regiment and moved behind the front into division reserve.

The battles for Mount Cosna exacted a terrific toll from the young troops. We had five hundred casualties inside of two weeks and sixty brave mountain soldiers lay in Rumanian soil. In spite of the fact that the major mission was not achieved and that we failed to destroy the enemy's southern flank, yet the mountain troops executed every assigned mission in a masterful way in the face of a hard-fighting, tenacious, and well-equipped enemy. I still look back on days as a commander of such troops with intense pride and joy.

A few weeks' leave in a Baltic resort were sufficient to put me back into shape.

Observations: In the defense on August 20, 1917, the main line of resistance was shifted to dense wooded terrain on the forward slope, in order to nullify the anticipated action of the Rumanian artillery. This was completely justified, for in the course of the fighting the enemy did not succeed in covering this concealed main battle line with artillery fire. The main defense positions were being prepared while the combat outposts withdrew fighting and the reserve companies were employed to dig covered communication trenches for the forward line. These trenches proved to be important for moving supplies of all kinds and evacuating the wounded under fire without losses, or at least with only slight ones. Afterward the reserves dug themselves in at the designated places.

The defensive fighting on August 20 required using the reserves at the frequently changing danger spots. Where danger threatened, the reserves had to occupy the main battlefield in depth. Reinforcing the front line proper with reserves was avoided as far as possible.

Chapter 10

THE FIRST DAY OF THE TOLMEIN OFFENSIVE

I: Approach March and Preparation for the Twelfth Battle of the Isonzo

Early in October I resumed command of my detachment. We were located in beautiful Carinthia to which the battalion had been sent via Macedonia. Replacements had made up the losses sustained in the Mount Cosna battles and the arrival of light machine guns increased the rifle companies' fire power. The short rest period was for thorough familiarization with the new weapon.

We had no idea as to the plans of the High Command. The Isonzo front?

Trieste had been the Italian operational objective since the start of the war in May, 1915, and ten battles along the lower reaches of the Isonzo had driven the Austrians slowly but continuously back. The results of the Sixth Isonzo Battle in August 1916 had been to allow the Italians to establish a bridgehead on the east bank near Görz (Gorizia) and to take the town itself.

Cadorna patterned the Eleventh Isonzo Battle (August 1917) on the Western Front model and 50 infantry divisions, supported by 5000 guns, attacked on the narrow front between Görz (Gorizia) and the sea. By fine fighting the worthy Austrian troops nullified the Italians' initial success, but in the second part of the battle the Italians crossed the middle reaches of the Isonzo and took the high plateau of Bainsizza where, by exerting their supreme efforts, our allies succeeded in halting the attack. This all-out attack lasted until the beginning of September when things quieted down and Cadorna began to get ready for the Twelfth Isonzo Battle. The newly won territory east of the middle reaches of the Isonzo materially improved the Italian prospects for the next battle and their objective, Trieste, was finally within reach. The Austrians did not feel equal to meeting this new attack and they were obliged to ask for German help. In spite of the tremendous expenditure of forces in the battles in the west (Flanders and Verdun), the German High Command sent an army consisting of seven battle-tried divisions. A combined German and Austrian offensive on the upper Isonzo front was to effect the desired relief. The objective was to throw the Italians back across the imperial boundary, and if possible back across the Tagliamento.

The Württemberg Mountain Battalion joined the newly created Fourteenth Army and was attached to the Alpine Corps. On October 18 we began our approach march to the front from assembly areas in the vicinity of Krainburg. In pitch-dark nights, often in streaming rains, Major Sprösser's Group (Württemberg Mountain Battalion and the Württemberg Mountain Howitzer Detachment No. 4) moved via Bischoflak, Salilog, and Podbordo to Kneza, which was reached on October 21. Because of enemy aerial reconnaissance each prescribed march objective had to be reached before daybreak at which time all men and animals had to be concealed in the most uncomfortable and inadequate accommodations imaginable. These night marches made great demands on the poorly fed troops.

My detachment consisted of three mountain companies and a machine-gun company, and I usually marched on foot with my staff at the head of the long column. Kneza was about five miles east of the battlefront near Tolmein. In the afternoon of October 21 Major Sprösser and his detachment commanders reconnoitered the assigned assembly area for the attack. This area was on the north slope of the Buzenika mountain (509) (1600) a mile south of Tolmein. The slope itself dropped sharply down to the Isonzo.

Thanks to their dominant positions, the Italians were able to bring many batteries into action for harassing fire against our rear areas. The Italians seemed to have plenty of ammunition. The assigned assembly area made it difficult to complete the battalion's (eleven companies) preparations. The slope itself was generally impassable and we had to make our preparations at the edges of some rock piles and in the few folds in the ground which fell sharply off in the direction of the Isonzo. It was disquieting that the enemy, from his very commanding positions on the Mrzli peak (1360) (4500) northwest of Tolmein, could get flank observation of the entire northern slope of the Buzenika mountain. In addition, we had to reckon with falling stones from the artillery bombardment on the steep slope. The battalion was obliged to remain in the assembly area for about thirty hours and during that time we often wondred as to the outcome of the operation.

We had to put up with all these unfavorable circumstances for the mass of troops being assembled was too large for the Tolmein basin. We returned to the battalion and passed under lively Italian harassing fire directed in particular on the St. Luzia and Baza-di-Modreja passes. Our knowledge of operational plans was considerably less than that of the Czech traitor who, on that day, deserted to the enemy, taking with him a complete set of maps and orders for the Tolmein operation.

The battalion moved into its final assembly area during the night of October 22. Huge searchlights from the Italian positions on the heights of the Kolovrat and Jeza illuminated our way. Heavy artillery fire frequently struck among us and the powerful and dazzling searchlight beams forced us to lie motionless for minutes at a time. As soon as they swept past, we hurried through the endangered area. During this advance we all received the impression of having come into the effective range of an exceptionally active and well equipped enemy. The pack animals had to be left behind on the east slope of Buzenika mountain. It was shortly after midnight that my detachment, heavily laden with machine guns and ammunition, reached its assembly area on the rock slope. It had been an exhausting climb. We dropped our loads and everyone rejoiced in having come through unscathed. Rest was out of the question for the remaining hours of darkness had to be used in digging and finding concealment. I assigned the companies their sectors. The staff and two rifle companies dug in on the western edge of the twenty to forty yard wide rocky slope which was bisected by a narrow path and had some defilade from the northwest. The two remaining companies occupied a narrow fold in the ground a hundred yards to the east. Everyone, officers and men, worked feverishly and dawn presented a lifeless slope. The soldiers tried to make up lost sleep in foxholes covered with shrubbery and branches.

But this peaceful quiet did not last long and the Italian heavy artillery soon paid us a visit and sent rocks rumbling past us on their way to the Isonzo. Sleep was again out of the question and we began to wonder if the enemy had perceived our preparation and was adjusting his fire. Heavy artillery fire on this roof-steep slope would have had a devastating effect.

The fire lasted a few minutes and then quieted down only to come to life fifteen minutes later in a different place. Then we had quiet for a while.

The Italian artillery shifted its main activity to the Isonzo valley. In the course of the day we observed the powerful effect of heavy caliber guns against the installations and approach roads near Tolmein. In contrast to the Italians, our artillery fired at rare intervals. I was much concerned regarding the welfare of the men entrusted to me and the day passed very slowly.

Our assembly area provided us with an excellent view of the forward Italian positions. They crossed the Isonzo a mile and a half west of Tolmein and then ran south of the Isonzo to the east edge of the Woltschach passing just east of St. Daniel. The positions, and especially the wire entanglements, seemed to be well constructed. The murky weather prevented our studying the other positions.

The second hostile position was said to cross the Isonzo in the region of Selisce—six miles northwest of Tolmein—and to run south of the Isonzo across the Hevnik to Jeza. The third and probably the strongest

Italian position had been installed on the heights south of the Isonzo in the line Matajur (1643), Mrzli peak (1356), Golobi, Kuk (1243), Hill 1192, Hill 1114 through Clabuzzaro, Mt. Hum. This latter position was known through aerial photographs. Isolated strong points were said to be located in the terrain between the individual positions.

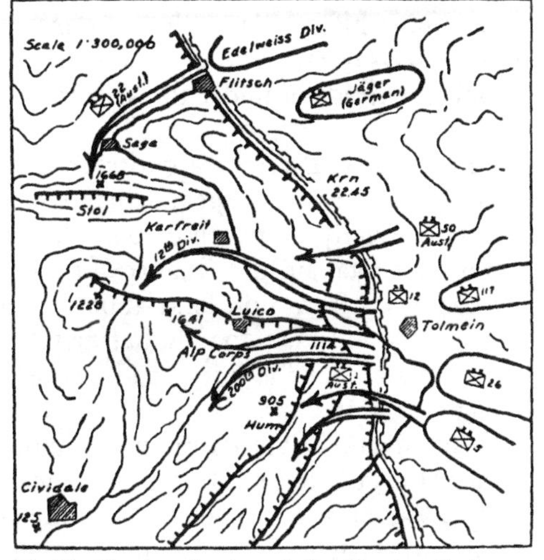

Sketch 34: Attack plan of the Fourteenth Army.

The forces of the Fourteenth Army were disposed as follows:

The Kraus Group ready at Flitsch (22d Imperial and Royal Infantry Division; Edelweiss Division; 55th Imperial and Royal Division; and the German Jäger Division) with the thrust point above Saga on the Stol.

The Stein Group near Tolmein and in the bridgehead position south of Tolmein (12th Infantry Division; Alpine Corps; 117th Infantry Division) was to execute the main attack. The 12th Division was to break through to Karfreit in the valley on both sides of the Isonzo; the Alpine Corps was to capture the positions on the heights south of the Isonzo (above all 114), Kuk, and Matajur.

Adjoining on the south, the Berrer group (200th and 26th Infantry Divisions) was disposed against Cividale via Jeza and St. Martino.

Still farther south the Scotti group (1st Imperial and Royal Division and 5th Infantry Division) was to take the positions south of Jeza, and later Globocak and Mount Hum.

In the Alpine Corps sector in the Bridgehead position south of the Isonzo Bavarian Infantry Life Guards and the 1st Jäger Regiment had relieved the Austrians in the front line.

The attack objective of the Life Guards: via Kovak, Hevnik, Hill 1114 (3675) Kolovrat Ridge, to the highway leading to Luico, Golobi, Matajur.

Attack objective of the 1st Jäger Regiment. Heights west of Woltschach, Knoll 732 (2415), Hill 1114 (3675) from the southeast.

The Württemberg Mountain Battalion had the mission of protecting the right flank of the Life Guards, of taking the hostile batteries, and of following the Life Guards to the Matajur.

Toward evening of October 23 the weather became misty, damp, and cold. The ration train arrived at dark; we soon satisfied our hunger and hunted up a hiding place where we might store up some sleep for the ensuing days of attack. A fine rain set in after midnight and forced us to pull our heads in under the shelter halves. Attack weather!

Observations: Even the approach march and the preparation for the attack at Tolmein made great demands on the troops. In strenuous night marches, usually in pouring rain, the Karawanken mountains were traversed, a total air line distance of sixty-three miles. By day the troops lay concealed from hostile aviation in very restricted shelter. Rations were scanty and monotonous, but despite all that, morale was high. In three years of war the troops had learned to endure hardships without losing their resiliency.

In the advance to the assembly area in the night of October 22 a reserve of machine-gun ammunition in belts was carried by the machine-gun company and units of the mountain companies. The Mount Cosna fighting had made the difficulty of ammunition supply in the mountains very clear.

Since strong enemy fire attacks had to be coped with in the assembly area, the troops dug in during the night and camouflaged the new installations carefully before daybreak.

It was impossible to supply the troops in the assembly area by day and hot food was brought up after dark.

II: The First Attack: Hevnik and Hill 1114

Our hitherto silent artillery began its preparation at 0200, October 24, 1917. It was a dark and rainy night and in no time a thousand gun muzzles were flashing on both sides of Tolmein. In the enemy territory an uninterrupted bursting and banging thundered and reechoed from the mountains as powerfully as the severest thunderstorm. We saw and heard this tremendous activity with amazement. The Italian searchlights tried vainly to pierce the rain, and the expected enemy interdiction fire on the area around Tolmein did not materialize for only a few hostile batteries answered the German fire. That was very reassuring and, half asleep, we retired to our cover and listened to the lessening of our own artillery fire.

At daybreak our fire increased in volume. Down by St. Daniel heavy shells were smashing positions and obstacles and occasionally their smoke obscured the hostile installations. The fire activity of our artillery and mortars became more and more violent. The hostile counterfire seemed to be rather weak.

Shortly after daybreak the Württemberg Mountain Battalion got under way and headed forward in a heavy rain which had greatly reduced the visibility. Following Sprösser's staff, which hurried ahead, the Rommel detachment descended the boulder strewn slope toward the Isonzo. Once below, we moved up behind the right wing of the Bavarian Infantry Life Guards just above the steep bank of the Isonzo.

A few shells struck on both sides of the long column of files without doing any damage. The column halted close to the front line. We were frozen and soaked to the skin and everyone hoped the jump-off would not be delayed. But the minutes passed slowly.

In the last quarter hour before the attack the fire increased to terrific violence. A profusion of bursting shells veiled the hostile positions a few hundred yards ahead of us in vapor and a gray pall of smoke. Low-hanging rainclouds covered the mountain tops of the Hevnik and the Kolovrat.

Shortly before 0800 the assault squad ahead of us left its positions and headed toward the enemy. The defenders, in the turmoil of fire, did not see or resist them and we took advantage of the released area in order to prepare for the attack.

0800! The artillery and mortar fire shifted in the direction of the enemy. In front of us the Life Guards rose to the attack. Following closely behind their right flank, we moved to the right front and gained the hostile positions around St. Daniel. The remnants of the garrison emerged from the ruins and hurried toward us with hands raised and fear-distorted faces. We hastened forward over the broad plain still separating us from the north slope of the Hevnik. To be sure, machine-gun fire from the eastern spurs of the Hevnik hampered our advance here and there; but our attack across the open surface kept moving.

While the Life Guards moved toward the east slope of the Hevnik, our objective was the northeast slope to which Major Sprösser and his staff hurried ahead of the soldiers who, hampered with heavy packs, machine guns, or ammunition, did not move forward so rapidly.

After reaching region 179, the wooded slope of the Hevnik protected our left flank from fire from the heights. (See sketch 35.)

The whole Rommel detachment reached the protecting slope. On orders from Major Sprösser, it moved up the footpath leading to Foni as advance guard of the Württemberg Mountain Battalion on the north slope of the Hevnik. Elements of the 1st Company under Technical Sergeant Seitzer formed the point. The remainder of the detachment followed with 150-yard intervals between units. A platoon of the 1st Machine-Gun Company followed the point, then the detachment staff, 1st Company, 2d Company, and the rest of the 1st Machine-Gun Com-

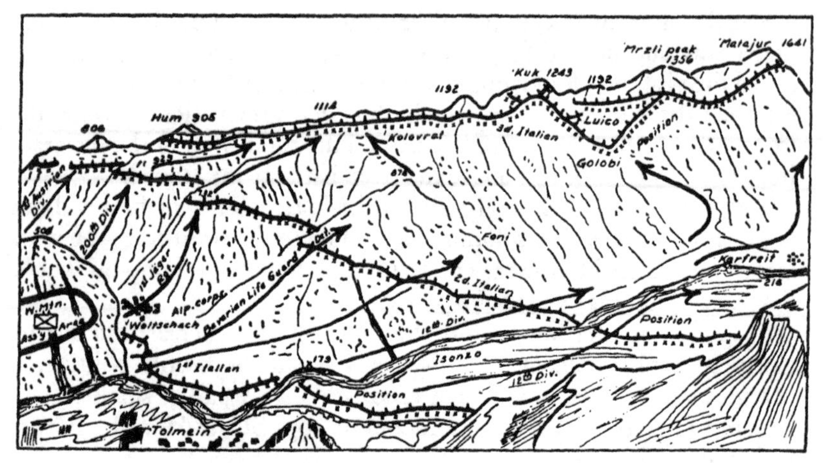

Sketch 35: Attack of the Fourteenth Army. View from the northeast.

pany. With Lieutenant Streicher, my new adjutant, I took my place in the column a few yards behind the point.

The footpath on which we climbed toward Foni was narrow and overgrown with bushes and gave little evidence of having been used by the enemy. The slope on both sides of the path was very steep and densely wooded. Autumn foliage was still on the trees. We had only a few yards visibility through the dense underbrush and seldom got a glimpse of the valley. A few deep rills led to the Isonzo. The impact of heavy German shells resounded dully up from the valley and also from the left rear where we supposed the Life Guards to be. The slope before us was unnaturally quiet and we expected to meet the enemy at any moment. None of our artillery was in position to help us in the mountain wood; we were completely on our own.

The point advanced with extreme care stopping frequently and listening in the wood, then moving forward again. But all this caution was to be of no avail for the enemy lay in ambush. When we had advanced a thousand yards east of Elevation 824 (2720), we were suddenly fired on by machine guns at close range. I received the report: "Enemy in front of us in developed position behind wire entanglements. Five men of the point are wounded."

Without artillery support an attack on both sides of the path along the roof-steep slope through dense underbrush and across obstacles against the very attentive and well dug in enemy appeared hopeless to me, or at least possibly only under heavy losses. Therefore I decided to try my luck elsewhere.

The existing point remained in contact with the enemy and I ordered another group of the 1st Company to act as the new point and to climb south through a stony draw about two hundred yards in front of the hostile position. My intention was to outflank the enemy from the left and from above. I informed Major Sprösser.

The ascent proved to be very difficult. Lieutenant Streicher and I followed forty yards behind the new point. Close behind us came the crew of a heavy machine gun carrying their disassembled gun on their shoulders.

At this moment a hundred-pound block of stone tumbled down on top of us. The draw was only ten feet wide and dodging was difficult and escape impossible. In the fraction of a second it was clear that whoever was hit by the boulder would be pulverized. We all pressed against the left wall of the fold. The rock zigzagged between us and on downhill, without even scratching a single man.

Happily, the supposition that the Italians were rolling stones down on us was false for the point had dislodged the stone.

Farther up the slope a rolling stone tore off my right heel strap and crushed my foot so badly that I required the help of two men to proceed during the next half hour. I was in agony.

Finally the steep draw was behind us. In pouring rain, wet to the skin, we climbed the slope through dense undergrowth, looking and listening intently in all directions.

The wood in front of us thinned out. My map showed that we must be half a mile east of Hill 824 (2720). We worked our way cautiously to the edge of the wood where we discovered a camouflaged path leading down the slope to the east. Beyond it, on the bare rising slope, we made out a continuous, well-wired position running uphill in the direction of Leihze peak. This hostile position appeared untenanted and no German artillery fire had come to bear on it. My decision was: A surprise attack after a brief heavy machine-gun preparation with our left flank along the edge of the wood. The situation was very reminiscent of the situations prior to the attacks on Mount Cosna on August 12 to 19, 1917.

Under the protection of a heavy machine-gun platoon disposed in concealed position in the bushes, I prepared the detachment for the attack in a small hollow in the woods sixty yards in front of the hostile obstacles. Thanks to the splendid combat discipline of the mountain troops, the movement was accomplished in the pouring rain without a sound. Far off the noise of battle resounded in the Isonzo valley, and somewhat nearer on the left rear on the ridge the Life Guards seemed to be fighting hard. Peace prevailed around us and on the surface of the meadow.

Now and then we saw a few men moving about in and to the rear of the hostile position—a sign that the enemy in front of us had no suspicion of our presence. A few German shells began to hit six hundred yards to our left rear. The hostile position in front of us must, to judge from its direction, connect with that position on both sides of the Foni road which we had encountered forty-five minutes before. My guess was that it was part of the second Italian position. Further soundless approach was impossible in the dense underbrush. The detachment was ready and I had to decide whether to attack or not. Sixty yards of underbrush separated us from the enemy wire and if he were on the alert I could not count on an easy victory.

The well camouflaged path along the wood's edge game me an idea. This path probably constituted the means of communication with the forward Italian line near St. Daniel or with the garrisons on the east slope of the Hevnik or the artillery observation outposts located there. Since our arrival it had not been used by the Italians. The path was winding and the camouflage on the south side gave such good concealment toward the upslope and in the direction of the Italian positions that it would be difficult to identify any troops using it. Without enemy interference we could move over the path and be in the enemy positions inside of thirty seconds. If we moved rapidly then we might capture the hostile garrison without firing a shot. A task for valiant man! If the enemy resisted, then I would have launched my prepared attack under the fire protection of the machine-gun company.

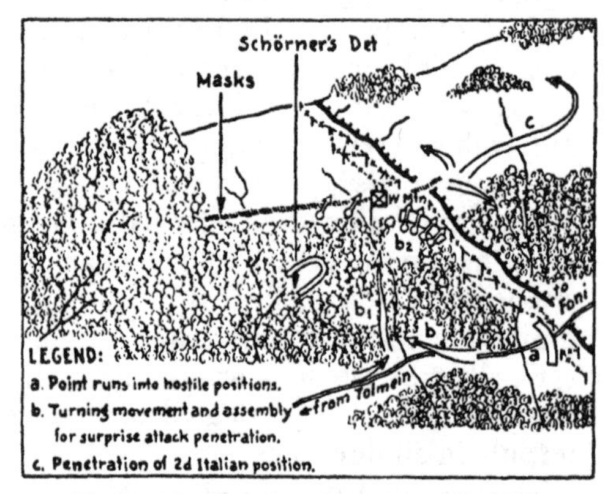

Sketch 36: Situation on October 24, 1917. View from the north.

I singled out Lance Corporal Kiefner, a veritable giant, gave him eight men, and told him to move down the path as if he and his men were Italians returning from up front, to penetrate into the hostile position and capture the garrison on both sides of the path. They were to do this with a minimum of shooting and hand grenade throwing. In case a battle developed they were assured of fire protection and support by the entire detachment. Kiefner understood, selected his companions, and a few minutes later took his squad down the camouflaged path. Their rhythmical steps died away and we began to speculate on their success.

We listened tensely, ready to attack or to exert steady fire. One shot would send three companies off to the attack. Again long, anxious minutes passed and we heard nothing but the steady rain on the trees. Then steps approached, and a soldier reported in a low voice: "The Kiefner scout squad has captured a hostile dugout and taken seventeen Italians and a machine gun. The garrison suspects nothing."

Thereupon I led the whole Rommel detachment (2d and 1st Companies, and 1st Machine-Gun Company) down the path and into the hostile position. The Schiellein detachment (3d and 6th Companies and 2d Machine-Gun Company), which had joined me shortly before Kiefner's successful breakthrough, followed us. Assault teams noiselessly widened the breach until we had fifty yards on either side of the path. Several dozen Italians, who had sought shelter in their dugouts from the streaming rain, were thus captured by the skillful mountain troops. Thanks to the heavy cover the enemy farther up the slope did not perceive the movement of the six companies.

I then had to decide whether I should roll up the hostile position or break through in the direction of Hevnik peak. I chose the latter. The elimination of the Italian positions was easy once we had possession of the peak. The farther we penetrated into the hostile zone of defense, the less prepared were the garrisons for our arrival, and the easier the fighting. I did not worry about contact to right and left. Six companies of the Württemberg Mountain Battalion were able to protect their own flanks. The attack order stated: "Without limiting the day's activities in space or time, continue the advance to the west, knowing that we have strong reserves with and behind us."

The 1st Machine-Gun Company was echeloned farther forward, for in case of a struggle I wanted to have a strong fire force right at hand. The heavy machine gunners, carrying loads of ninety pounds, determined the speed of ascent. This gigantic accomplishment can only be understood by one who has climbed around in the high mountains with a similar load and under similar weather conditions.

Our thousand-yard column worked its way forward in the pouring rain, moving from bush to bush, climbing up concealed in hollows and draws, and seizing one position after another. There was no organized resistance and we usually took a hostile position from the rear. Those who did not surrender upon our surprise appearance, fled head over heels into the lower woods, leaving their weapons behind. We did not fire on this fleeing enemy for fear of alarming the garrisons of positions located still higher up.

During the advance we were repeatedly endangered by our own strong artillery fire. We did not give light signals to shift the fire forward, since they would have alerted the hostile garrisons. One man of the detachment was wounded by a rock dislodged by a heavy German shell.

Our booty included a 210mm battery which had been gassed and whose crew had disappeared without leaving a trace. Mountains of shells were piled by the giant guns and the dugouts and ammunition dumps blasted in the rock were undamaged. Three hundred feet up hill we came upon a medium battery whose guns were located in absolutely shellproof rock casemates equipped with small loopholes. Here, too, the crew had vanished.

At 1100 we reached the ridge running from Hevnik peak toward the east where we made contact with units of the 3d Battalion of the Life Guards and accompanied them for a way along the ridge toward the Hevnik peak, which was under heavy German fire. While the Life Guards rested and waited for the artillery to shift its fire, I turned off with my companies toward the north slope of the Hevnik and reached the peak at noon without encountering any resistance. We saw numerous groups of scattered Italians and took a number of them prisoner.

The rain had stopped and the clouds hanging low overhead began to move so that we had occasional glimpses of Hill 1114 (3675) and the Kolovrat Ridge from whence heavy artillery fire was being directed at the Hevnik. Apparently we had been detected by Italian observers in front of Hill 1114. To avoid needless losses I moved both detachments out of the endangered area in a northerly direction and, in accordance with our mission, let them clean up hostile artillery nests between the Hevnik and Foni. Reconnaissance detachments secured the southern slope of the Hevnik and the Nahrad Saddle which lay three hundred yards southwest of the peak. We marked all our booty with chalk and the bag had risen to seventeen pieces including twelve heavy caliber guns. Italian preserves and a prepared meal appeased our great hunger.

Elements of the Life Guards arrived in the Nahrad Saddle around 1530 and I joined them with my two assembled detachments. A half hour later the 3d Battalion of the Life Guards (3 rifle companies) began to climb the camouflaged main track leading to Hill 1114 via 1066 (3520). Bearing in mind the fact that our misson was to protect their right flank, I followed them with my six mountain companies. Rommel's detachment led, followed by Schiellein's.

Lieutenant Streicher and I marched at the head of my column. The weather had cleared and the Kolovrat Ridge, Hill 1114, and the ridge running from Hill 1114 to Jeza were sharply outlined. For the moment no enemy hindered our ascent. Around 1700 the leading Life Guard company was fired on as it approached Hill 1066, and two of the companies took cover under the cliffs to the east of the path.

I ordered the Rommel detachment to march up under cover on the right of the path to the level of the 3d Battalion companies of the second line, and then Lieutenant Streicher and I reconnoitered the area in vicinity of Hill 1066 where we met elements of the 12th Life Guard engaged in a fire fight with a strong enemy located in a series of superimposed positions on the hill six hundred yards northwest of Hill 1114 and on Hill 1114 itself. These positions dominated the area and appeared to be well wired. Italians were also in position to the right of the road running past the 12th Company's right flank. I quickly moved the 1st Company under Lieutenant Triebig forward and ordered it to clear the enemy from the positions on the right of the road in the area southwest of Hill 1066. The company executed this mission with neatness and dispatch, and we managed to take the positions at no cost to ourselves. Our prisoner bag showed seven officers and 150 men.

Meanwhile in accordance with my orders, the 2d Company and 1st Machine-Gun Company cleared up the trenches, dugouts, and observation posts west of Hill 1066. The Schiellein detachment came up and went into reserve a hundred yards northwest of Hill 1066 and just below the rocky crest which we had finished clearing.

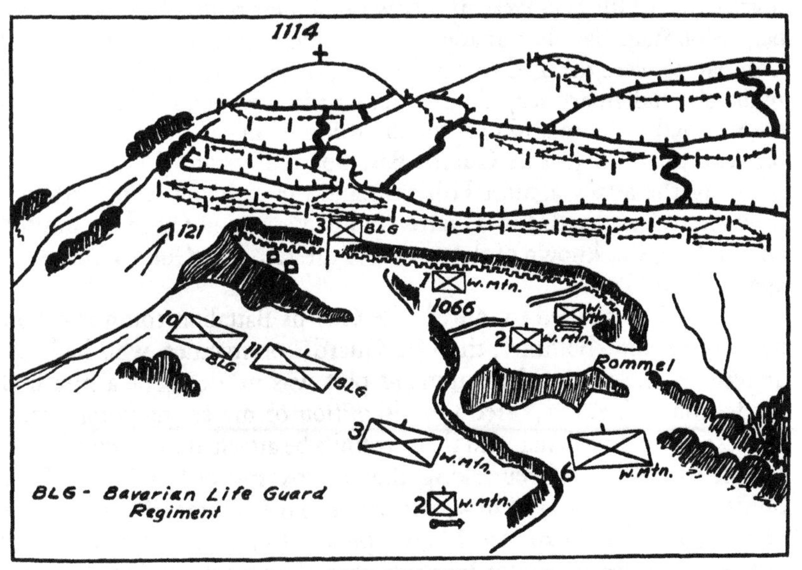

Sketch 37: In front of Hill 1114. View from the northeast.

Lieutenant Streicher and I headed for the right wing of the 12th Life Guard Company. We thought their positions superior to ours for close observation of Hill 1066, and we also wanted to establish closer liaison with the 3d Life Guard Battalion. We got in the front line where, some fifty yards short of Hill 1066, we met several 3d Battalion officers who showed us a scout squad attempting to creep up to the nearest hostile position by moving through the draw leading toward the saddle between Hill 1114 and the hill six hundred yards northwest of Hill 1114. The scouts did not enjoy too favorable prospects; for the enemy, who was still unharmed, was covering the bare slopes in front of his wire with additional bursts of machine-gun fire. This local enemy garrison appeared to be on its toes and was not inclined to surrender any more ground.

The officers of the 3d Life Guards Battalion, Lieutenant Streicher and I were all of the opinion that the strong positions on Hill 1114 and on the hill six hundred yards to the northwest could only be taken in co-operation with the artillery. Up to that time neither hill had been subjected to artillery fire. Using my field glasses, I made a careful study of the details of the hostile positions, although a machine gun emplaced somewhere on Hill 1114 frequently obliged me to abandon my observations and dive for cover.

Darkness fell slowly and all attempts of the 1st Company to take additional sectors of the hostile positions on the hill six hundred yards northwest of Hill 1114 were unsuccessful. My elements of the Württemberg Mountain Battalion made ready for the night, and the 1st and 2d Companies were detailed to furnish combat reconnaissance during the night. An Italian artillery observation post located behind the 1st Company served as command post for the Rommel detachment. Lieutenant Streicher, various 3d Life Guards Battalion officers and I discussed our plans for the attack against Hill 1114 and the Kolovrat Ridge. At that time the 10th and 11th Life Guards Companies had not been committed and nothing was known of the success of the 12th Life Guards Company against Hill 1114.

At 1900 I was called to the 3d Life Guards Battalion command post by Major Count Bothmer, the Life Guards commander, who had just arrived on the scene. His command post was in a dugout a hundred yards from mine. I reported the disposition of my six mountain companies and he then demanded that my units be attached to his command. I took the liberty of remarking that I took my orders from Major Sprösser, who, as far as I knew, was senior to the commander of the Life Guards, and that I expected Major Sprösser to arrive at my command post at any moment. Count Bothmer replied by forbidding me to move any part of my detachment to the west or against Hill 1114, saying that this was work for the Life Guards only. He graciously permitted the units of the Württemberg Mountain Battalion to follow his Life Guards on October 25 and to occupy and secure the height and then to follow him in second line. I told him that I would inform my commander of his actions. Then I was dismissed.

I was none too happy on my return to the command post. Fighting in second line did not appeal to us mountain troops at all, and I searched for ways and means of achieving complete freedom of action for my force. My final conclusion was that I would have to wait until Major Sprösser arrived.

At 2100, Lieutenant Autenrieth, the battalion quartermaster, arrived at our command post. He had been directed to us from the 12th Life Guards Company via the 3d Life Guards Battalion command post, where he had been present at a discussion of the attack planned for October 25. This attack was to be made against Kolovrat Ridge and included artillery support. He told me that Major Sprösser had continued the attack on Foni with Wahrenberger's detachment and had broken in there just prior to dark. Lieutenant Autenrieth also reported that the 12th Infantry Division had made excellent progress in the Isonzo valley. I described the situation at Hill 1114 and our relationship with the Life Guards, urged him to report this to Major Sprösser as soon as possible, and to ask him to come to Hill 1066 with or without the Wahrenberger detachment before daybreak and thus restore my detachment's freedom of movement. Lieutenant Autenrieth gladly accepted this job which was very difficult in the pitch-black night through terrain not completely cleared of the enemy and started off for the Group command post.

Wet clothes and cold wind made the night most unpleasant for the elements of the Württemberg Mountain Battalion on Hill 1066. Night patrols of the front line companies brought in a few dozen more prisoners, who were captured in front of the hostile obstacles. However, no patrols succeeded in getting through the obstacles into the foremost hostile position. The Italian sentries were extremely watchful and were quick to use hand grenades and machine-gun fire.

Late in the evening the 3d Battalion of Life Guards informed us that the reserve companies located north of Hill 1066 had been committed against the left of the northeast slope but that they had failed to establish contact with the 1st Jäger Regiment which was attacking across Hill 732 (2415). We were not told that Lieutenant Schörner's company (12th Life Guard Company) had taken Hill 1114.

Half asleep on a hard bed I thought about the continuation of the attack. A frontal attack? Such an attack required artillery support and this support would not be available until after daylight on October 25. Moreover the Life Guards did not want the Württemberg Mountain Battalion to participate in an attack from our newly won positions against the strong Kolovrat defenses. An alternative to attacking with artillery support was to hit the Italians in a sector which, up to that time, had remained unassailed. The sector I selected for such an operation was the third Italian position lying some eleven hundred yards east or southeast of Hill 1114. To the west of Hill 1114 this position crossed the bare, terrace-like summits of the Kolovrat Ridge which mounted up to the Kuk. A successful breakthrough west of Hill 1114 would have an effect on the positions lower down. This situation was one offering attractive possibilities to the aggressive officers and men of the Württemberg Mountain Battalion. Hill 1114 dominated the enemy positions to the southeast. A breakthrough down there would have had little effect on the situation at Hill 1114 and was out of the question for the Württemberg Mountain Battalion, which was on the right of the Life Guards but had been denied all operational activity by Major Count Bothmer. The night passed quietly broken only by a short hand-grenade skirmish.

The scout squads sent out in the early morning hours against the hostile positions fared no better than the night patrols and were repulsed by alert Italian sentries. The 3d Battalion of Life Guards gave us no indication that the situation had changed during the night. It was still pitch-black at 0500 when Major Sprösser arrived at my command post. The rest of the Württemberg Mountain Battalion (4th Company, 3d Machine-Gun Company) followed close behind him. I described the situation on Hill 1114, our relationship with the Life Guards, and my plan of attack for whose execution I requested the use of four rifle and two machine-gun companies.

Major Sprösser agreed to my plan of operation against the third Italian position, but only gave me two rifle and one machine-gun companies, though promising additional support in case of success. While I made the arrangements for the departure of my new formation, Major Sprösser reached an understanding with the Life Guards' commander, who arrived at my command post.

Observations: The first Italian position at St. Daniel consisted of a continuous trench in the front line with numerous dugouts, shelters, and strong wire entanglements. Individual machine-gun nests and strong points were located in the zone between the first and second positions. Front-line camouflage was deficient, while the installations between the first and second lines were barely perceptible.

The German artillery preparation demolished the front line and all but annihilated its garrison. The few machine-gun nests in the zone between lines which were not destroyed by the preparatory fire were unable to stop the attack which was launched on a broad front. If the Italian had had numerous machine-gun nests in the zone between the first and second positions, then the German attack would probably have been stopped. Gargantuan artillery preparation is required to demolish a modern defensive position which is laid out in great depth.

My detachment's point lost five men upon encountering the second Italian position on a narrow path along a steep and wooded slope. An increased interval between men would have reduced losses. In Rumania the Cossack points rode at intervals of more than two hundred yards when moving over open terrain. If something happened to the first man, the next one reported it. An infantry point must do likewise and the point commander must combat the herd impulse to close up.

While the Italian garrison of the second position on the way to Foni proved to be very wary, the garrison of the same position half a mile southeast was not sufficiently on the alert. It is not enough to have watchful sentries in the position; the forward area must be constantly surveyed by patrols, especially in bad weather and in irregular and covered terrain.

Order of battle at daybreak October 25: The Kraus Group attacking in the Flitsch basin reached Saga on the evening of October 24 in the drive down the valley. It began to attack the Stol (5400 feet altitude) on the morning of October 25.

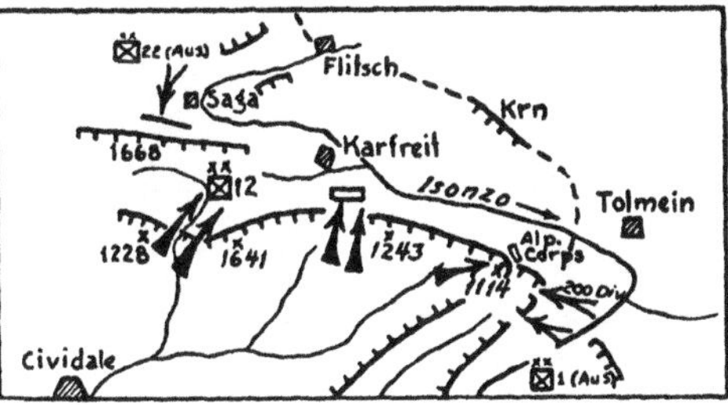

Sketch 38: The situation at daybreak, Oct. 25, 1917.

In the Isonzo valley the 12th Division—favored by the rainy, murky weather, which nullified hostile fire effect from the mountains into the valley—had advanced on October 24 through Idersko and Karfreit to the Natisone valley near Creda and Robic. The Eichholz Group (2 battalions 1 artillery platoon) had branched off toward the Luico Pass. On the morning of October 25 weak units of the 12th Division (Schuieber's company) climbed the northern spurs of the Matajur massif and Eichholz's Group was engaged in hard fighting against greatly superior Italian forces.

In the Alpine Corps, the Bavarian Infantry Life Guards and the Württemberg Mountain Battalion were fighting at the cornerstone of the third Italian position on Hill 1114. Schörner's company (12th Life Guards) held the peak proper, but the Italians held their surrounding positions and were attempting to regain their lost positions by counterattacks. The 1st Jäger Regiment was still fighting for the second Italian position in the region of Hill 732.

The 3d Jäger Regiment of the 200th Division had taken Jeza, and the 4th Jäger Regiment was fighting for the second Italian position west of Hill 497.

The Scotti Group had, with the 1st Imperial and Royal Division, taken the first and second Italian positions and reached the line of Ostry-Kras-Pusno-Srednje-Avska.

Summary: The third Italian position on the powerful heights south of the Isonzo (Matajur, Mrzli peak, Golobi, Kuk, Hill 1192, Hill 1114, La Cime, Mt. Hum) remained with the exception of small segments on Hill 1114, in Italian hands. Their garrison was fresh, and they had ample reserves. The position had not suffered from German artillery fire.

Chapter 11

THE SECOND DAY OF THE TOLMEIN OFFENSIVE

I: Surprise Breakthrough To the Kolovrat Position

Just before dawn on October 25, 1917, I left the western part of Hill 1066's rocky summit with the 2d Rifle and 1st Machine-Gun Companies and started down to the northwest through a steep and narrow draw. Our initial objective was a clump of bushes 150 feet below our starting point. The movement was not executed without loss, for the enemy noticed us and sprayed the column with machine-gun fire. Nevertheless we reached the bushes where we were joined by the 3d Company of my detachment. At this moment violent action broke out on Hill 1114.

Before we started out I had instructed the Company commander as to the proposed scheme of maneuver. My intention was to move west until I was from an eighth to a quarter of a mile below the Kolovrat positions, a move which would place me a mile and a quarter away from the fighting on Hill 1114. Once in position on the steep slope I would await a favorable opportunity to launch a surprise attack against the third hostile position. Success for the undertaking rested on our not being seen by the Italians as we moved across the slope.

Lieutenant Ludwig's 2d Company sent its point out, and I guided it by hand signals. My staff (adjutant, runners, and wire detail) and I followed thirty yards behind the point. Fifty yards behind us in column of files came the 2d Rifle, 1st Machine-Gun and 3d Rifle Companies. Part of the wire detail was busy establishing communications with Major Sprösser's command post on Hill 1066.

Being on the move was extremely pleasant after a cold night spent in wet clothes. Italian preserves had to replace our morning coffee. As the day grew lighter, the combat on Hill 1114 and in the vicinity of Hill 1066 increased in violence. We moved away from this noise of battle and noiselessly worked our way from bush to bush and from slope to slope. At first the overgrown terrain allowed us to move along just six hundred feet below the hostile positions; then obstacles appeared on the bare knobs of the long Kolovrat Ridge and forced us to execute time- and energy-consuming detours down the slope. Up above among the hostile obstacles, perhaps even in front of them, many watchful sentries' eyes surveyed the slopes on which we moved. Were any of them to see us and give the alarm then the whole success of my enterprise was jeopardized if not irretrievably ruined.

We halted and made the necessary personal reconnaissance whenever we felt our security menaced. So much depended on finding the correct way. We cautiously crossed several deeply cut ravines and continued our advance over a grassy slope. The entire column had to be concealed from enemy observation and this was difficult since we could only guess about how the terrain appeared to those above us. We also had to guess regarding the enemy's position which, because of the heavy obstacles, appeared very strong. The higher we went the scarcer the bushes, and our covered approach was finally reduced to employing the narrow rills which furrowed the slope. An hour after leaving Hill 1066, we had covered about twenty-two hundred air yards from that place; and we had not been fired on at any time. From Hill 1114, we still heard the sound of battle, which led us to believe that the Life Guards had renewed their attack.

The fortified knobs of Kolovrat Ridge sparkled above us in the morning sun and it had all indications of a beautiful fall day. Deep silence prevailed around us. The point was working its way past a few clumps of bushes and into a hollow some two hundred yards from the enemy wire. I was considering whether and where I could cross the bare, sharp ridge lying about a hundred yards ahead of us when I heard a slight noise behind me. Looking back, I saw some mountain troops of the 2d Company disappearing into a large group of bushes below the path traversed by the point.

What was going on? The soldiers at the head of the 2d Company had discovered some Italians asleep in a clump of bushes down the slope. Inside of a few minutes they had routed out an Italian combat outpost of forty men and two machine guns. Not a shot, not a loud word was heard. To be sure, a few hostile sentries fled downhill as fast as their legs could carry them; but fortunately in their excitement they forgot to warn the garrison of the positions above by shots or shouts. I made certain that no one tried to shoot them as they fled.

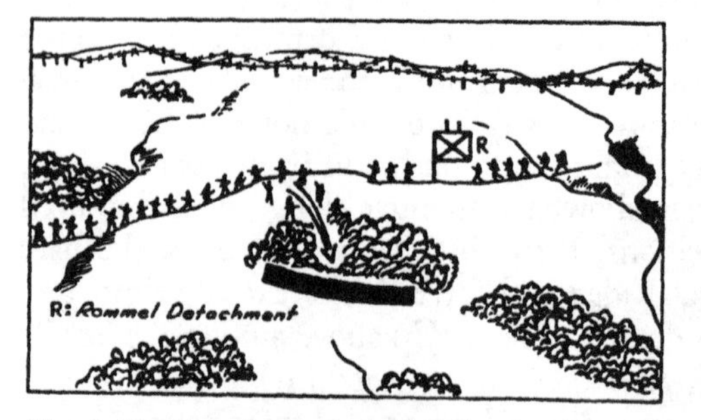

Sketch 39: Surprising the Italian Combat Outpost, October 25, 1917.

This enemy combat outpost apparently had the mission of securing the positions on the Kolovrat Ridge from surprise from the direction of the Isonzo valley. There were probably additional outposts three hundred feet or so below us, but they evidently had their attention fixed on the Isonzo valley and it never occurred to them that we might advance westward from Hill 1066.

The noiseless elimination of the principal security element for the Kolovrat position greatly improved our chance for success in approaching the enemy obstacles. Finally the deepest part of the draw, where the point had halted, was invisible to the Italians from their position on the heights. This series of circumstances led me to decide to launch my breakthrough at that particular point.

The prisoners were sent to the end of the column, and I ordered the point to proceed up the draw to within a hundred yards of the hostile obstacles. The tops of the wire stakes were just visible to me. The point provided security for the assembly area. With the greatest care and caution I moved my companies into the draw and prepared for the attack. The area itself was very small and I had a high troop density. I issued my instructions to the commanders and then we moved off behind the point which was a hundred yards closer to the enemy. The slopes were very steep and definitely arched.

Nothing stirred in the position ahead of us, but fighting was still in progress on Hill 1114.

My adjutant, Lieutenant Streicher, offered to reconnoiter the obstacles in front of us to discover their strength or weaknesses, and, if necessary, to cut gaps. I gave him five men of the 2d Company and a light machine gun, instructing him to use his firearms as weapons of last resort. Streicher crept up forward with his men. Lieutenant Ludwig maintained contact with him by means of a few riflemen.

Meanwhile the telephone squad had established contact with Major Sprösser's command post (near 1066), and I reported the course of events and gave him my decision to break into the hostile Kolovrat position immediately at a point half a mile east of Hill 1192. I also requested and was promised the speedy dispatch of support in case of success. Major Sprösser had followed our entire progress with his glasses from his command post. He told me that the situation in front of Hill 1114 had changed only to the extent that strong Italian forces were attacking the Life Guards. An attack by the Life Guards with artillery support had failed to materialize.

I had just put down the telephone and was biting into an Italian white roll when a brief report came from Streicher: "Scout squad broke through, took guns and prisoners." Complete silence reigned in the enemy position and not a shot had been fired. With maximum speed I proceeded with the execution of my breakthrough plan and got my entire detachment underway. A second's delay might have snatched away the victory which lay in our grasp.

It required all our strength to climb up out of the draw and across the steep slope. In a few moments the hostile obstacles were reached and passed and then we moved across the hostile position. The long barrels of a heavy Italian battery loomed before us and in its vicinity Streicher's men were cleaning out some dugouts. A few dozen Italian prisoners stood near the guns. Lieutenant Streicher reported that he surprised the gun crews at their ablutions.

We were in a narrow saddle. The bare knolls of the Kolovrat Ridge were covered with numerous earthworks, and communication trenches leading to the strongly developed position along the north slope were also visible. The main road leading to Crai through Luico—Kuk—Hill 1114 was only a hundred yards south of our position. The road itself was well camouflaged against ground and air observation.

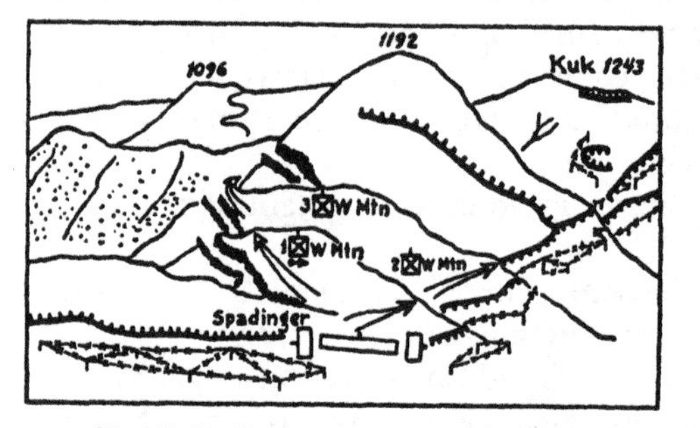

Sketch 40: Penetration of the position on Kolovrat Ridge. View from the northeast.

A third of Rommel's detachment had reached the saddle. The men were gasping from the effects of running up the steep slope. The Kolovrat garrison was still unaware of our entry into their positions. Was it still asleep? To judge from the number of prisoners already captured in the fifty-yard-wide saddle, the position was heavily occupied. Seconds separated us from our fate.

I ordered:

"Rommel detachment blocks to the east and extends to the west.

"Technical Sergeant Spadinger with 1 machine-gun squad from the 2d Company blocks the north shoulders of the hostile position, bars the ridge road, and covers the rear of the Rommel detachment advancing to the west.

"Lieutenant Ludwig and the 2d Company break open the hostile position on the western part of the north slope. Shooting is to be avoided as long as possible.

"3d rifle and 1st Machine-Gun Companies accompany me down the ridge road to the west. Lieutenant Streicher and his detachment assumes responsibility for the security of the advance.

"Advance as rapidly as possible!"

All units of the detachment went about their jobs with neatness and dispatch. Under the efficient Lieutenant Ludwig, assault squads of the 2d Company raced from dugout to dugout and from sentry post to sentry post. Most of the hostile garrison was encountered in the dugouts. One mountain soldier was enough to supervise the evacuation, disarming and falling in of a hostile dugout garrison. In the sentry posts the sentries were still watching the valley where they had a beautiful early morning picture of the Isonzo and its twin 6500-foot peaks gleaming in the sunlight.

The sudden apparition of a mountain soldier behind a sentry was enough to paralyze them with fear, and they came no nearer to giving the alarm than did their brethern of the combat outpost who we had taken just before the attack. The number of prisoners grew rapidly and soon reached into the hundreds.

The bulk of the detachment also made good progress down the ridge road. It was fortunate that the camouflage concealed us from the enemy on the heights to the east and west. We captured several gun positions which had been blasted right into the rock wall. Our sudden appearance in the quiet of the morning, far from the noise of battle near Hill 1114, completely disconcerted the garrison. My initial objective was to surprise any massed reserve formations, and I also wanted to reach a position from which I could assist the 2d Company in overcoming any resistance it might meet.

Events took a different turn.

About ten to fifteen minutes elapsed since our penetration of the Kolovrat position, and the 3d Company's point moved over the ridge road and was approaching the saddle 300 yards east of Hill 1192 when it was fired on from all sides.

The Streicher scout squad, which had already reached the saddle three hundred yards east of Hill 1192, was subjected to machine-gun fire from the southern slope of Hill 1192 and was soon hard pressed by Italian infantry advancing from the southeast slope of Hill 1192, trying to push north across the ridge road. The scout squad gave way on the northeast slope of Hill 1192.

The advance of the 3d Rifle and the 1st Machine-Gun Companies down the ridge road was stopped by heavy machine-gun fire from Hill 1192. The machine-gun units were hurriedly emplaced but they were unable to obtain any fire advantage. Machine-gun fire was striking through the camouflage and down along the left side of the road. This made an attack up one side of the road extremely difficult, for we would have moved across the steep and unprotected southern slopes of Kolovrat Ridge. Shortly after this I heard increased sounds of combat from the right front where I imagined the 2d Company to be. Hand grenades burst, followed by the lively carbine fire of the mountain troops. Every man seemed to be in the firing line.

I saw nothing and it was impossible to go to the bare knoll to the right of the road without drawing heavy machine-gun fire from Hill 1192. I began to wonder whether the 2d Company was holding the enemy. It had only 80 carbines and 6 light machine guns! If the 2nd Company was overrun, then the enemy would have had little trouble regaining his lost positions on the north slope and would have cut off the remaining units of the detachment and released the prisoners. The volume of fire told me that we were opposed by strong enemy forces. A few minutes had sufficed to change the situation completely in our disfavor and to make it very serious. It became a question of holding against a superior enemy the parts of the Kolovrat position won during our rapid advance. The most urgent need was to block the road to the west and rush to the help of the threatened 2d Company. Numerous enemy machine guns to the east and west swept the bare prominences which lay on the shortest road to the 2d Company. An attack astride the road toward the west against Hill 1192 would have been covered by the same enemy fire and would have small chance of succeeding. I arrived at a different solution.

A machine-gun platoon already engaged in fire against Hill 1192 and some riflemen of the 3d Company were given the mission of blocking the ridge road to the west. With the rest of the 3d Rifle and the machine-gun companies I hurried down the road to the saddle half a mile east of Hill 1192. The heavy camouflage prevented the enemy in the east and west from observing this movement and subjecting it to aimed fire. The occasional sweeping of the camouflage with searching fire did little to hamper our movement, and we managed to reach the saddle.

Here the capable Spadinger and his eight men were holding the Italian garrison of the position to the east in check. In moving past I reinforced him with two additional squads. We moved at the double toward the west through the Italian positions which had previously been mopped up by the 2d Company. A hundred and sixty yards farther on we came upon two mountain soldiers who were guarding about a thousand prisoners in the area between the position and the wire entanglements. I ordered them to move the prisoners down the slope below the wire entanglements immediately, and left the details to them. They carried out their job! The Italian machine-gun fire from the east and west grazing the heights accelerated the movement of the prisoners.

Some hundred yards ahead of us the noise of battle rose to great violence near the 2d Company. Hand grenades burst, machine guns fired steadily and carbines were delivering rapid fire. I demanded the utmost speed of the companies following me. I surveyed the situation from a knoll four hundred yards east of Hill 1192.

The 2d Company held some sections of trench on the northeast slope and was encircled from the west, south and east by fivefold superiority, an entire Italian reserve battalion. The foremost enemy units were massed a scant fifty yards away. The wide and high Italian obstacles lay in rear of the 2d Company making retreat to the north slope impossible. The troops defended themselves desperately against the powerful enemy mass; only their unbroken rapid fire prevented an enemy attack. If the enemy ventured to attack in spite of the fire, then the little group would have been crushed. I did not think it proper to commit my troops as soon as they arrived under enemy fire.

My estimate of the situation was that the 2d Company could be relieved only by a surprise attack by the entire detachment, this attack to be against the enemy flank and rear. Under such conditions I believed that the superior combat capabilities of the mountain soldier would prevail.

The first units rushed breathlessly through the deep trenches, followed by the leading soldiers of the machine-gun company with disassembled weapons. Only a few words were necessary to give the commanders the situation and their missions. We assembled the 3d Company in a shallow draw to the left of the trench. A heavy machine-gun crew hurriedly set up its weapon in a hollow on the right and reported itself ready for combat. The crew of another heavy machine-gun came up panting and the 3d Company was in the draw ready to attack.

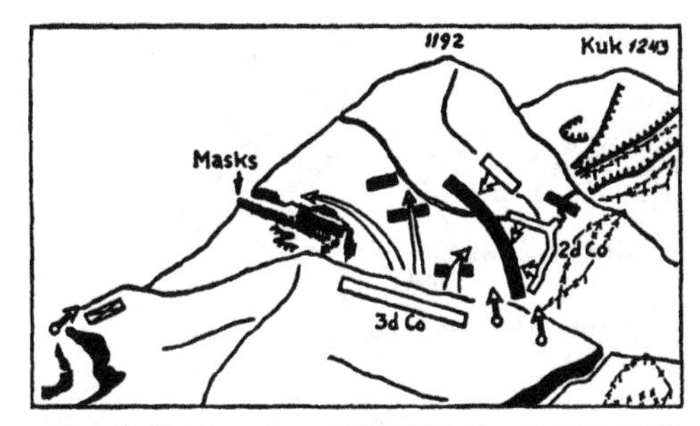

Sketch 41: Attack on Hill 1192, October 25, 1917
View from the east.

I could not wait for the second heavy machine gun to be set up. The dense enemy masses, urged on by their officers, were rising out of their trenches a hundred yards away and advancing on the cornered 2d Company. I gave the attack signal to the 3d Rifle and the 1st Machine-Gun Companies. While the first heavy machine gun opened up on the enemy with a steady fire, the second heavy machine gun shortly came into action and the mountain troops on the left charged the enemy flank and rear with savage resolution. Loud shouts resounded. The surprise blow on the flank and rear hit home. The Italians halted their attack against the 2d Company and tried to turn and face the 3d Company. But the 2d Company came out of its trench and assailed the right. Attacked on two sides and pressed into a narrow space, the enemy laid down his arms. Not until we were within a few yards did the Italian officers defend themselves with pistols. Then they, too, were overpowered. I had to interfere to save them from the fury of the mountain soldiers. An entire battalion with 12 officers and over 500 men surrendered in the saddle three hundred yards northeast of Hill 1192. This increased our prisoner bag on Kolovrat position to 1,500. We took the peak and south slope of Hill 1192 and captured another heavy Italian battery.

Our great joy over our success was saddened by our heavy losses. Aside from several wounded, two particularly valiant fighters, Lance Corporal Kiefner (2d Company), who distinguished himself so gallantly the day before on the Hevnik, and Technical Sergeant Kneule (3d Company), sacrificed their young lives in close combat.

By 0195 the Rommel detachment was in undisputed possession of a half mile sector of the Kolovrat position including Hill 1192 and extending to the east. Thus a wide breach had been driven into the main hostile position. The first hostile counterattack of local reserves resulted in their elimination. I had to reckon on further attempts of the enemy to regain what he had lost. Let the Italians come! We mountain troops were unaccustomed to surrender what we had gained by hard fighting.

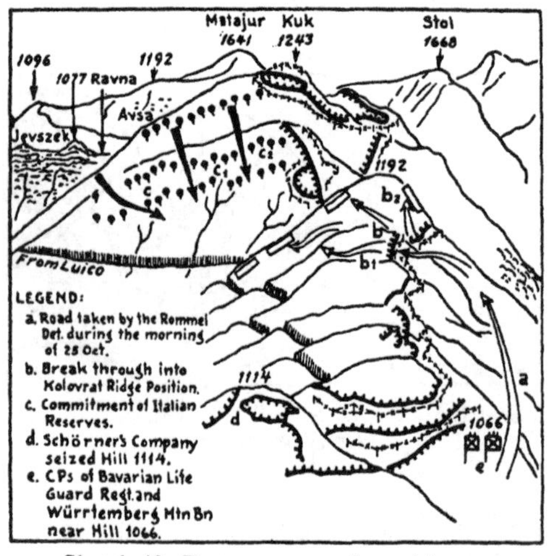

Sketch 42: Enemy reserves in position on Mount Kuk.

From the west, southeast, and east the enemy covered the heights held by us with machine-gun fire. Also Italian artillery groups on Mt. Hum and to the west had not missed the break-through on the Kolovrat and the battle of Hill 1192. Their heavy shells forced us to the north slope, which offered some cover.

My available reserves were insufficient to permit an immediate continuation of the attack. We had to hold our gains until support arrived. The 2d Company and half the MG company occupied Hill 1192 with front to the west. Spadinger with a platoon blocked the penetration's shoulder half a mile to the east. I kept the 3d and half the MG company in reserve in the newly won positions on the northeast slope of Hill 1192.

Then I studied the situation round about from the summit of Hill 1192. At first glance the western front in the direction of Kuk appeared to be the most dangerous one. Aside from dozens of machine guns, which fired on us from terrace-like and predominantly superior positions on the northeast slope of the Kuk, strong reserves were visible on the highest part and on the southeast slope. Soon several waves of skirmish lines began to move toward us across the broad east slopes of the Kuk. I estimated their strength at one or two battalions. In the south Mount Hum teemed like an ant hill and strong enemy artillery was firing. Lively automotive traffic in both directions was moving over the ridge road leading from Cividale over Mount Hum. On both sides of the road closed enemy formations were moving toward the battlefront. In the east we overlooked the whole Kolovrat Ridge, which descended gradually to Hill 1114. Strong hostile masses were distinguishable on the south and southwest slopes of Hill 1114 where the Italians appeared to be attacking. Long motor columns were bringing Italian reserves up from Crai and were detrucking them on the west slope of Hill 1114. Also enemy forces were soon moving along the ridge road and across the hills. They were moving up against us from the east. By all appearances the enemy intended to attack us simultaneously from two sides.

Observations: The surprise breakthrough into the Kolovrat position on October 25, 1917, was successful because the Italians did not watch the area in front of their third position sharply enough, a mistake which the Rumanians made time and again on Mount Cosna.

Also the garrison of the position itself was not prepared for action. Everyone thought it was safe from all danger, being a mile and a half from the current combat zone on Hill 1114. Thus initially the mountain troops had an easy time of it.

The counterattack of the Italian reserve battalion, undertaken with great energy, was stopped by the fire of the weak 2d Company, but it would probably have led to the annihilation of the 2d Company had we not succeeded in attacking the strongly massed Italian battalion on flank and rear at the decisive moment. It would have been a mistake to conduct this attack with too small forces or to limit it only to a fire attack from the flank.

After a successful breakthrough into the Kolovrat position (0915, October 25, 1917,) the order of battle was as follows:

The Kraus Group with the 1st Imperial Rifle Regiment was attacking in three columns from Saga. Their objective was the line Stol (1668)—Hill 1450.

In the Stein Group, the 12th Division with the 63d Infantry Regiment was near Robic and Creda as on the previous evening, and had repulsed hostile advance guards. The Schrieber detachment reported that it was a hundred yards north of the summit of Mount Matajur. (It was apparently a question of the Mount Della Colonna.) The Eichholz Group had been attacked by superior Italian forces from the Luico pass and was fending itself in tenacious fighting with this enemy. It held positions north of Golobi.

In the Alpine Corps, the Rommel detachment had succeeded in breaking into the Kolovrat position. The penetration extended half a mile east of Hill 1192. The bulk of the Württemberg Mountain Battalion was marching from Hill 1066 to Hill 1192. The Life Guards held the position near Hill 1114 reached in the evening of October 24 against violent Italian attacks. The 1st Jäger Regiment had taken Hill 732 and was advancing against Slemen Chapel.

In the 200th Division, the 3d Jäger Regiment had taken Hill 942 west of Jeza.

The Scotti Group: In the 1st Imperial and Royal Division, the 7th Mountain Brigade was attacking the Globocak.

II: Attack Against Kuk. The Barring Luico-Savogna Valley And Opening Of The Luico Pass

Contrary to my expectations, the enemy suspended his advance, which had been carried forward in several waves across the east slopes of the Kuk. Did he intend only to block off, or was he making additional attack preparations? The former proved to be the case, for the hostile riflemen began to dig in on three lines one above the other on the east slopes of the Kuk in conjunction with the north slope positions. An attack by these forces, supported by numerous machine guns in superior positions, would have worried me considerably. The enemy's adoption of defense and the consequent cessation of fighting was most welcome to me, for I knew that Major Sprösser was on the way toward Hill 1192 with the bulk of the Württemberg Mountain Battalion.

As soon as more forces arrived on Hill 1192, I intended to attack the enemy on the Kuk. He was to be given as little time as possible for digging in, for it would be increasingly difficult to dislodge him if he were permitted to anchor himself firmly. It was essential to use the time for thorough preparation of the planned attack.

In order to maintain surprise I refrained from hampering his entrenchment with fire. The rocky ground made digging difficult. Since the battalion staff was on the march, I made my report to Alpine Corps Headquarters via the switchboard on Hill 1066 and gave them the results of our operations and told them I intended to continue the attack upon arrival of reinforcements on Hill 1192. Furthermore, I explained my plan of attack against Kuk to Captain Meyr, the General Staff Officer of the Alpine Corps, and requested the support of the heavy batteries for the attack. My request was acceded to and, in a few minutes, I was connected with the fire-control officer of an artillery unit near Tolmein. We agreed that a preparation would be fired by the heavy batteries between 1115 and 1145 against the broad east slope of the Kuk and the positions on the northeast slope. With artillery support for the attack assured, I counted a great deal on the effect of heavy shells in the rocky terrain because of falling stones.

It was necessary to prepare the infantry fire support. For this purpose I emplaced the light machine guns of the 2d Company and the whole 1st Machine-Gun Company on the north and south slopes of Hill 1192. Their positions were concealed from the enemy on the Kuk. I planned to use weak assault teams in the attack and the automatic weapons had the mission of pinning down the enemy force on the Kuk. Targets were designated for each gun.

Major Sprösser arrived in the saddle just east of Hill 1192 at 1030 bringing with him the 4th and 6th Rifle Companies, and the 2d and 3d Machine-Gun Companies. I brought him up to date as to the situation and the preparations already made for the attack against the Kuk, and requested the attachment of the forces necessary for the attack. After studying the hostile position, Major Sprösser gave the 6th Company under Lieutenant Hohl the job of rolling up the hostile positions on the Kolovrat Ridge in the direction of Hill 1114. My plan of attack against the Kuk was approved and was given the 4th Rifle and 2d and 3 Machine-Gun Companies in addition to the 2nd, 3d Rifle and 1st Machine-Gun Companies. I soon completed my attack preparations.

By 1100 the whole fire detachment (six light machine guns, 2d Company, and 1st Machine-Gun Company) under Lieutenant Ludwig was in concealed position on the north and south slopes of Hill 1192, ready to open fire on the Kuk garrison. An assault team of the 2d Company consisting of two squads, was in the position on the north slope of Hill 1192 and an assault team of the 3d Company, of equal strength, was ready to advance on the south slope. The task of these assault squads was to take the saddle between Kuk and Hill 1192 upon opening of fire under strong fire support by artillery and machine guns, and then to advance against the Kuk garrison as far as possible along the position on the north slope or through the draw on the southeast slope. I wanted to feel out the hostile position with these assault teams. The 3d and 4th Rifle Companies and 2d and 3d Machine-Gun Companies were in concealed reserve positions in the saddle just east of Hill 1192. I intended to commit them on the north or south slope depending on the success of the initial assault team.

Just before the beginning of the attack the head of the Life Guards arrived in the saddle east of Hill 1192. Previously the 2d Life Guards Battalion had tried, after a vain wait for artillery support, to attack the positions on the Kolovrat Ridge from Hill 1114 but their attack had been stopped by heavy defensive fire from Italian positions five hundred yards northwest of Hill 1114. Then the Life Guards moved along the path taken by the Württemberg Mountain Battalion on the north slopes of the Kolovrat Ridge below the still tenaciously held positions between Hill 1114 and the saddle half a mile east of Hill 1192. Here they met the 1,500 prisoners taken by the Rommel detachment, who were being moved off by a few mountain soldiers.

Punctually at 1115 the first heavy shells from the Tolmein basin roared up and burst in the midst of the newly constructed Italian lines on the east slope of the Kuk. Falling stone rumbled downhill. A good beginning for the attack! Now the machine-gun fire units on Hill 1192 went into action and the assault teams on the north and south slopes of the height got under way. Under great tension I followed their progress with my field glasses.

The enemy on the Kuk answered our machine-gun fire and a regular machine-gun duel took place between our garrison on Hill 1192 and the Italians on the Kuk. An ear-splitting uproar! Shell after shell struck the enemy positions. The resulting splinter effect and the stone avalanche caused intense nervous strain among the enemy garrisons. The hostile artillery units on Mount Hum chimed in from the left flank but failed to find remunerative targets on the south slope of Hill 1192, for machine guns were well dug in and the artillery fire bothered them but little. Down on the right, on the north slope, hand grenades burst as the assault team from Ludwig's company fought its way along the hostile position. The Italian garrison clung tenaciously to every nook and cranny, and our troops made slow progress even though they were attacking downhill.

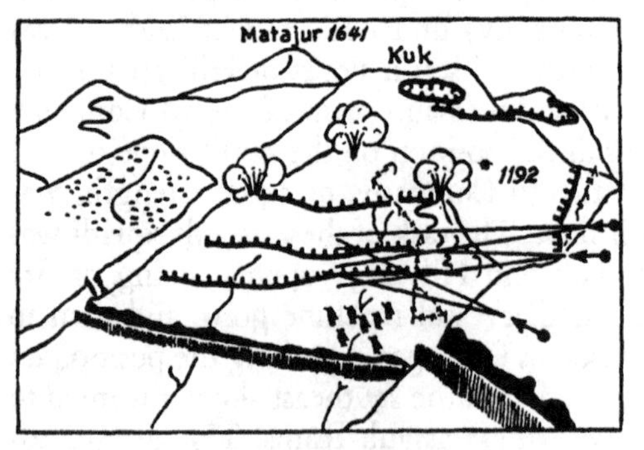

Sketch 43: The attack on Hill 1192, October 25, 1917. View from the east.

It was a different story

on the south slope of Hill 1192, where the assault team from the 3d Company vaulted out of the gun emplacements along the camouflaged road and quickly disappeared from view. Friendly and hostile machine-gun fire whistled over its head but, by hugging the camouflage, the team managed to reach the saddle between Hill 1192 and the Kuk without being subjected to aimed fire. From this location the assault team of the 3d Company began to climb toward the enemy on the summit regardless of our own artillery fire and the resulting stone avalanches. My observer followed all its movements.

Our artillery was well placed and shell after shell smashed into the Italian lines. Our volume of machine-gun fire increased where the assault squad of the 3d Company was closest to the enemy. Soon the assault team was within hand-grenade range of the enemy front line, and some of our men began to wave their handkerchiefs at the enemy who was exposed to our fire almost without any cover. The trick worked and we saw the first deserters run down from their position.

This was the moment to launch my main attack. I had four companies available. My orders to the assembled company commanders were:

"The south assault team is climbing the Kuk, and taking prisoners. The Rommel detachment attacks with its four companies up the northeast slope of the Kuk. The 3d Machine-Gun Company, 4th and 3d Rifle Companies, and 2d Machine-Gun Company follow the detachment staff down the camouflaged ridge road. Speed is at a premium.

"The fire support element on Hill 1192 gives maximum fire support and follows as soon as the situation permits."

We rushed forward down the camouflaged road. Had the enemy on the Kuk been attentive, he should have observed our movement. To all appearances, however, his entire attention was riveted on our machine guns on Hill 1192 and on the local hand-grenade fight. The expenditure of ammunition on both sides was heavy but only a few shots strayed toward the ridge road. Under these circumstances it did not take us long to reach the saddle between Hill 1192 and the Kuk which was defiladed from the Italians on the Kuk. The entire detachment followed at the double in column of files.

While this maneuver was in progress the assault team had increased its prisoner bag to one hundred. A report from the rear stated that elements of the Life Guards would join us in our advance down the ridge road. This addition gave command of a force which exceeded a regiment in strength and which was strung out behind me for a distance of two miles. The question arose regarding the advisability of extending our objective.

For the next quarter of an hour our artillery and machine-gun fire pinned the enemy to his positions on the east of the Kuk. The 3d Company assault team gathered in all Italians pried loose by our fire. The camouflaged ridge road, which swung around the south slope of the Kuk, and its garrison offered an attractive avenue for an all out attack. I had visions of cutting off the Kuk garrison. To be sure, I had to figure on fighting additional strong reserves on the south slope and on the fact that the defender could use considerable forces and charge down the steep slope. But on the other hand, I knew that no task was too difficult for my mountain troops who had been thoroughly tested on many fields, and I did not hesitate to advance. The attack continued.

My objective was Ravna, a small mountain village on the southwest slope of the Kuk. I tore down the road with the head of my detachment. The few riflemen of the point were followed by Grav's machine-gun company. The men, gasping for breath and dripping with sweat, carried their heavy machine guns on their shoulders. It was tough going and they had been carrying their guns since the opening attack, but they all knew that they must exact their last ounce of physical strength.

The ridge road, still very well camouflaged, ran down in the direction of Ravna. It had been blasted into the almost-bare steep slope of the Kuk, and the hostile garrison on the slope above was unable to see what was going on along the road. Their entire attention was still focused on the battle with Hill 1192. On the other hand, we had a small field of vision on the road and the numerous turns never gave fields of view greater than a hundred yards in depth. Vertical stone walls out of vision on the right and the camouflage on the left restricted our view in that direction. This narrow field of sight acted to our advantage.

At frequent intervals, sometimes only a few yards apart, we ran into an unsuspecting enemy standing or marching down the road. He never had an opportunity to use his arms before being taken. A sign to disarm and a gesture toward the east were enough to start the weaponless Italians marching down our column toward Hill 1192. They were all paralyzed by our sudden appearance.

We rushed on past battery positions, supply trains, closed hostile infantry formations without being stopped or even fired on. To the right rear and up the slope the fire fight between Hill 1192 and the Kuk garrison was still in progress and a few stray shots hummed past us high overhead. The Italians on the Kuk were still waiting for the Germans to start the usual infantry attack on a broad front across the slopes of Hill 1192.

The masks along the left side of the road stopped just short of Ravna widening our field of view. The net picture consisted of long bare slopes broken by occasional rows of small bushes. We pondered whether Italian reserves were concealed in or behind these bushes. The first houses of Ravna were about three hundred yards ahead of us. To the left, down the steep slope, were several farms, and behind them the wooded Knoll 1077. Once more we increased our pace to the utmost and reached Ravna without being fired on.

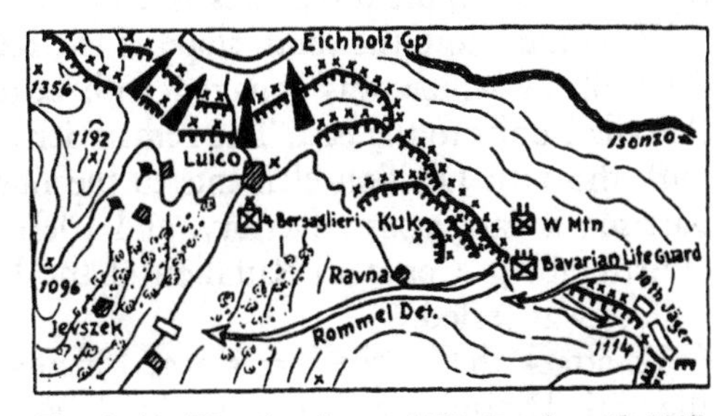

Sketch 44: The situation at 1200, October 25, 1917.

It was noon and the hot sun beat down on the south slope. No wonder that the Ravna garrison, thinking itself to be far from the front, failed to discover us until we were storming through the few houses and barns. The terrified Italians scattered and fled head over heels down into the Luico and Topolo valleys; their pack animals stampeded. To our great amazement, not a shot was fired and the southern slope of the Kuk appeared to be lifeless. The reserves up there had been committed against Hill 1192.

The last of the Ravna garrison, probably some pack-animal detachment, disappeared beyond the small knoll just west of the village in the direction of Luico. We followed right on their heels. I reached the knoll with the foremost soldiers of my detachment and had a splendid view, particularly to the west.

Down on the right was the mountain village of Luico in the saddle between the Kuk and the Mrzli. The town and the fairly-large neighboring cantonment were full of Italian troops. The usual peaceful rear area activity was going on in and around Luico. The Luico-Savogna was crowded with two-way vehicular traffic. Among other things, a heavy horse-drawn battery was moving at a walk from Luico to the south. The sound of heavy fighting came from north of the town, and I imagined them to come from the attack of the 12th Infantry Division.

[It was Eichholz's Group—three battalions strong, which was resisting the counterattack of strong Italian forces, planned against Karfreit through Iderslo and intended to hit the flank and rear of the 12th Infantry Division, which had moved forward in the valley north of the Matajur. Translator.]

On the other side of Luico the winding Matajur road ran up and over the partially-wooded eastern slopes of the Mrzli and Mount Cragonza. Little traffic was observed there. Italian artillery units were in position near Ausa and Perabi and were firing on units of the 12th Division near Golobi.

The remaining units of my detachment followed me at top speed, for I did not wish to lose momentum in Ravna but wanted to act rapidly and in a decisive direction. There was no time for lengthy deliberations. I rapidly weighed the three lines of action open to me.

We could climb the south slope of the Kuk and capture its garrison, the bulk of which was engaged with other Württemberg Mountain Battalion units to the east. The remainder of the garrison was engaged with the 12th Division elements to the north. I did not consider this garrison a dangerous adversary, and I left it to rearward units of the Württemberg Mountain Battalion of the Life Guards. In my opinion its fate was sealed.

To attack the hostile forces near Luico and open the Luico Pass for the 12th Division was an attractive line of action. My two machine-gun companies could give excellent fire support from superior positions. The possibilities of approaching the massed hostile forces around Luico were excellent and the attack would come as a surprise. However, it failed to insure the annihilation or capture of the enemy around Luico, because the rugged and wooded terrain on the east slope of the Mrzli offered the enemy the possibility of evacuating the pass without suffering excessive losses. I rejected this line of action and decided: to cut off the hostile forces around Luico by blocking the Luico-Savogna valley and the Matajur road on Mount Cragonza (1096). The wooded slopes on both sides of the Luico-Savogna valley favored this line of action

Sketch 45: The attack against the Luico—Savogna Road.

for they permitted our reaching the valley near Polava, before the hostile forces around Luico suspected our presence. With the valley and the road blocked and the Alpine Corps at Luico, the encircled enemy could not have avoided annihilation or capture.

Was the detachment too dispersed? I had been unable to keep all units in sight along the camouflaged road along the southern slope of the Kuk and it was more than likely that the rapid pace of the advance had extended the column more than was desirable. Waiting for the units to close up was now out of the question, for our success depended on valuable seconds.

From Ravna I turned sharply to the southwest with the foremost units of the detachment and moved toward the Luico-Savogna valley in the vicinity of Polava on the wooded west slope of Hill 1077. I sent runners back to Ravna with instructions to send all companies of the detachment in the direction of Polava.

As we hurried by, we snatched eggs and grapes from the baskets of the captured pack animals. We moved on at the double! I cautiously bypassed Knoll 976 up on the left, for I was unable to determine whether or not it was occupied by hostile forces. I did not want to be stalled. As on the Kolovrat Ridge a few hours before, I chose the way through clumps of bushes and small woods, for we had to move unobserved by the enemy in Luico and on Hill 976. Movement downhill was easy across the soft meadows. We decidedly wanted to capture the heavy battery moving off from Luico in the direction of Savogna. We quickly approached the valley floor.

The head of my detachment reached the valley a mile and a half southwest of Luico at 1230. The sudden apparition of the leading soldiers, among them Lieutenants Grau, Streicher, Wahrenberger and myself, who suddenly rose from the bushes a hundred yards east of the road, petrified a scared group of Italian soldiers who were moving unsuspectingly along, partly on foot and partly awheel. They were totally unprepared to encounter the enemy two miles behind the front at Golobi and they fled at top speed into the bushes to the side of the road, probably expecting to be fired on at any moment. But nothing was farther from our minds than that.

We reached the road and began to dig in at a point where the road made two sharp turns. All enemy wire lines were cut immediately. The 4th Company and 3d Machine-Gun Company, which had arrived, were disposed in the bushes and undergrowth on the slopes on both sides of the valley so that, though invisible, they commanded the valley by fire far to the north and south.

Unfortunately it turned out that we had lost contact with the remaining companies shortly after passing through Ravna, that is while we were on the west slope of Hill 1077. That was a hard blow for I needed at least two or three more companies to execute my planned advance against Mount Cragonza and to block off the Matajur road. I sent Lieutenant

Walz back with the mission of bringing up the remaining companies as soon as possible and of informing Major Sprösser of our accomplishments and of our future plans.

Meanwhile, to our great astonishment, Italian traffic started up again on the Luico-Savogna road. From north and south single soldiers and vehicles came unsuspectingly toward us. They were politely received at the sharp curves of the road by a few mountain soldiers and taken prisoner. Everyone was having fun and there was no shooting. Great care was taken that the movement of the vehicles did not slacken on the curves. While a few mountain troops took care of the drivers and escorts, others seized the reins of the horses or mules and drove the teams to a previously designated parking place. Soon we were having trouble handling all the traffic that came from both directions. In order to make room, the vehicles had to be unhitched and moved close together. The captured horses and mules were put in a small ravine immediately behind our barricade. Soon we had more than a hundred prisoners and fifty vehicles. Business was booming.

The contents of the various vehicles offered us starved warriors unexpected delicacies. Chocolate, eggs, preserves, grapes, wine and white bread were unpacked and distributed. The worthy troopers on the slopes on either side were served first. Soon all efforts and battles of the past hours were forgotten. Morale two miles behind the enemy front was wonderful!

This happy state of affairs was disturbed by a sentry's warning. An Italian automobile was approaching at high speed from the south. A wagon was quickly dragged across the road, but a machine gunner who fancied the victim already escaping, fired at fifty yards, against my express command. The automobile stopped suddenly in a cloud of dust; the driver and three officers jumped out and surrendered with the exception of one officer who reached the bushes below the road and escaped. A fourth soldier lay mortally wounded in the car. They were officers of a higher staff in Savogna who, perturbed by the interruption of telephone connection with the front, wanted to find out personally regarding the state of the fighting. The automobile proved to be undamaged; its former driver drove it to the parking place.

About an hour had passed since we blocked the road and there was still no trace of the remainder of the detachment. We did not hear sounds of heavy combat either in the direction of Luico or Kuk. We hoped the hostile front had not closed behind us. In this case we would have been obliged to cut our way through to our own lines.

A new report of the sentry on the east side of the valley focused our attention to the north. A very long column of Italian infantry was approaching from Luico. Under the impression that they were far behind the front, the head of the column came marching up to us in a most peaceful manner. It had no security.

Alarm! Get everything ready for combat. I expected to be fighting within the next few minutes with 150 mountain troops against great numerical superiority. But our position was strong and our machine guns commanded the valley for a considerable distance. The closer I let the enemy advance toward our blockade, the less chance he had of deploying properly and attacking with his larger force. I ordered all to withhold fire until I gave the signal with my whistle.

The head of the hostile column was within three hundred yards of our blockade. In order to avoid unnecessary bloodshed, I sent Deputy Officer Stahl toward the enemy as intermediary with a white armband. He was to order the enemy to lay down his weapons without a struggle, pointing out our possession of the slopes on both sides of the road. While he hurried toward the column, Lieutenants Grau, Wahrenberger, Streicher and I stepped out in front of the bend in the road. By waving handkerchiefs we intended to emphasize Stahl's words.

Stahl reached the head of the hostile column. Officers rushed forward, snatched off his pistol and field glasses which he had not discarded because of haste, and took him prisoner. He scarcely got a word in edgewise. Our waving did not help. The Italian officers ordered the leading groups to fire on us. We quickly disappeared around the bend. Then my whistle loosed a hail of fire from both slopes on the enemy column that swept the road clear in a few seconds. While the enemy was taking cover, Stahl succeeded in escaping and hurried back to us.

Since we had to be very careful with ammunition, I ordered the men to cease firing after a minute. The answering fire was weak. By waving a handkerchief I again demanded surrender. Too soon! The enemy made use of the pause in fire to rush out of the bushes in deployed formation. At the same time several machine guns opened fire on us from the slope just west of the road. Then it did become evident who was the better shot. Our fire from concealed and very superior positions had excellent

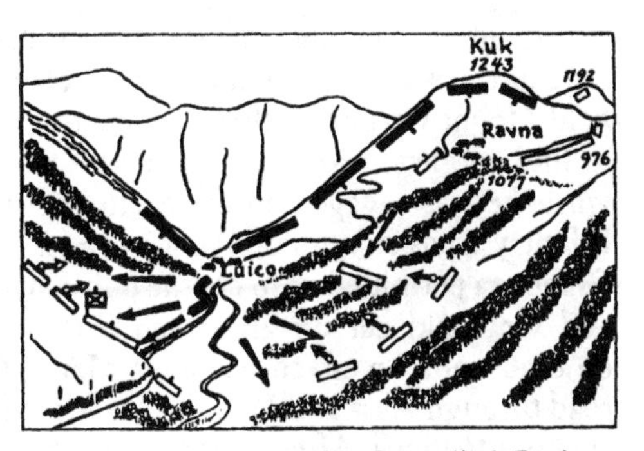

Sketch 46: The fight with Bersaglieri Regiment near Luico. View from the south.

effect on the still densely massed enemy. After five minutes of fire fight I once again demanded their surrender. Again in vain! Again his forward units charged us during the pause in firing. They were about ninety yards away.

Finally, after ten minutes of very violent firing the enemy conceded that he was beaten and gave the sign of surrender. Thereupon we ceased fire. Fifty officers and 2,000 men of the 4th Bersaglieri Brigade laid down their weapons on the valley road and moved toward us. I entrusted the capable Deputy Officer Stahl with the job of gathering and moving the prisoners through La Glava and Hill 1077 to Ravna. I gave him a few riflemen as escort.

We were reinforced by the 3d Company, which had also intervened during the final phase of the fight with the Bersaglieri from the slopes east of the valley. Sounds of violent combat had been coming from the direction of Luico for some time. In order to clarify the situation there, I armed the captured automobile with a heavy machine gun and traveled in the direction of Luico. It took quite some time to get by the Italian weapons and equipment which littered the road for nearly three quarters of a mile. Just south of Luico I met the heavy Italian battery observed from Ravna; its team lay dead in the road. When I arrived in Luico itself toward 1530, the other units of the Württemberg Mountain Battalion under Major Sprösser and the 2d Battalion of the Life Guards had just reached Luico and the valley to the south. They had attacked from Ravna. I met Major Sprösser at the south entrance of the town just as units of the 2d Life Guards Battalion were driving the enemy down the Matajur road in the direction of Avsa.

I suggested to Major Sprösser that with all available units of the Württemberg Mountain Battalion I climb toward Mount Cragonza on the shortest way across country from Polava and capture that peak. If we held Mount Cragonza, then the enemy on the Mrzli peak would have to find another road to the south and we could attack him in the rear while he was engaged on the north and northeast by units of the 12th Infantry Division and the Alpine Corps. Besides that, control of Mount Cragonza permitted us to cut the only ridge road leading to the Matajur and we could cut off all Italian batteries emplaced near or moving on that road. An advance against Mount Cragonza along the Matajur road through Avsa and Perati seemed less desirable to me. What was the enemy situation? After giving up Luico pass strong Italian units were moving, in fairly-good order, along the Matajur road to the east slopes of the Mrzli-Cragonza hill mass. Presumably they intended to occupy prepared rear positions there. On the Matajur road a weak rear guard was adequate to hold off the pursuer. This gave him time to reassemble his units and for deliberate occupation of the prepared positions. It was also assumed that the positions on both sides of the Matajur road were occupied. These considerations caused me to suggest climbing Mount Cragonza by the shortest possible route.

Major Sprösser agreed and gave me the units of the Württemberg Mountain Battalion in and south of Luico (2d, 3d, and 4th Companies, 1st, 2d, and 3d Machine-Gun Companies, and the Signal Company). At the same time the Gössler detachment (1st, 5th, and 6th Companies of the Mountain Machine-Gun Detachments 204 and 205) was ordered to move toward Luico at the disposal of Major Sprösser. Major Sprösser himself went to brigade headquarters in the Italian auto captured at Polava, in order to report regarding our previous battles and to secure artillery support for the anticipated actions.

Observations: The decision of the Italian commander on the Kuk to stop the German breakthrough in the Kolovrat position by committing his numerous reserves for defense in several lines on the east slope of the Kuk was incorrect. He gave the Rommel detachment the urgently needed respite (for organization of the defense, reassembly, bringing up of support). It would have been more advantageous to use these forces to recapture Hill 1192. The necessary fire support could have been given from the numerous positions on the north slope of the Kuk. If the hostile command had succeeded in getting an attack going from the east against the Rommel detachment, the latter would have been in a very difficult position.

Further, it was not profitable to locate the three positions on the steep, bare and stony east slope of the Kuk (forward slope). In hours of work the Italian soldiers barely succeeded in denting the ground even though their work was not disturbed by any harassing fire. Reverse slope positions on the west slope of Hill 1192 would have been much more favorable for the enemy since they would have been out of reach of our artillery and machine guns.

Furthermore, the enemy delayed in blocking the ridge road on the south slope of the Kuk and in covering the bare slopes below the ridge road with fire.

At the start of the attack against the Kuk, two or three Italian battalions opposed the Rommel detachment with numerous machine guns in commanding positions, in part well developed, in part hastily installed. The detachment first attacked only with two assault teams of 16 men each under the fire support of one machine-gun company. Six light machine guns and two heavy batteries felt out the possibilities of approaching the enemy and finally used the main body to encircle the entire Kuk garrison, which was captured during the later hours of the morning by assault units from the Württemberg Mountain Battalion and a company of the Bavarian Infantry Life Guards.

In the attack the effect of the machine-gun and heavy-artillery fire against the hastily entrenched enemy proved to be especially strong. In many places they were unable to stand up under this severe nervous strain. This fire would have had little effect had the Italians been properly entrenched.

Our own machine-gun fire from Hill 1192 proved to be a magnet and attracted the entire attention of the Italian fire units, thus permitting our initial assault team and then the entire detachment to reach the eastern slope of the Kuk without suffering losses by means of the camouflaged road which was open to enemy observation.

In Ravna the communications within the Rommel detachment broke down because a machine-gun company commander began to round up some of the captured mules. The result was that I reached the valley near Polava with only a third of my forces, could only block the Luico-Savogna valley, and had to give up blockading the Matajur road in the region of Mount Cragonza. To be sure, the units which dallied at Ravna later took part in the attack against the enemy at Luico, but our success would have been even greater had Mount Cragonza come into our possession on October 25. Doctrine: If an attack breaks into the defensive zone or a breakthrough succeeds, the reserves must stay up with the leading units and must not be diverted by taking booty, etc. On such occasions the most rapid pace is required of all rearward units.

A regiment of the 4th Bersaglieri Brigade in march column unexpectedly bumped into our road block in the narrow valley. Even though the leading units were pinned down by fire, the rearward units could have mastered the situation by attacking on the slopes to the east or west. Clear thinking and vigorous command were lacking here.

In the afternoon of October 25, 1917, the order of battle was:

Kraus Group: The 1st Imperial Rifle Regiment was attacking Stol from Saga. The 2d Battalion had taken the Hum and the 1st Battalion had seized the Pvrihum. The 43d Brigade was climbing Hill 1450. The assault company of the 3d Imperial Jäger Regiment was taking Mount Caal, the 13th Company of the 3d Jägers was attacking Tanamen Pass.

Stein Group: 63d Infantry Regiment of the 12th Division was advancing down the Natisone valley as far as the border two miles south of Robic and was driving all Italian reinforcements back. The Italian positions on the north slopes of the Matajur were not being attacked. Eichholz's Group was still engaged a mile north of Golobi and was gaining ground slowly, taking Golobi at 1700 and reaching Luico at 1800 only to find the town already in possession of the Bavarian Infantry Life Guards and the rear units of the Württemberg Mountain Battalion. In the Alpine Corps, the Kuk garrison was captured at 1406 by units of the Württemberg Mountain Battalion and a company of the Life Guards. At the same time the 6th Company of the Württemberg Mountain Battalion rolled up the Kolovrat Ridge from Hill 1110 to Hill 1114. After encircling the Kuk and cutting off the Luico-Sagovna valley, the Rommel detachment captured considerable units of the 4th Italian Bersaglieri Brigade in a fight near Polava. The bulk of the Württemberg Mountain Battalion and the 2d Life Guards Battalion had taken Luico by attacking from Ravna. The 1st and 10th Jäger Battalions were fighting on the south slopes of Hill 1114 and, in the course of the afternoon, captured Hill 1044 and all of Hill 1114. In the 200th Division the 3d Jäger Regiment was fighting south of Hill 1114 in the vicinity of Crai and, at 1800, the 4th Jäger Regiment took La Cima half a mile north of Hill 1114.

Scotti Group: The 8th Grenadier Regiment crossed the Judrio in an attack from Ravna against Mount Hum. The 2d Mountain Brigade took the Cicer and the 22d Mountain Brigade took St. Paul.

The results were: On October 25 the powerful third line of Italian positions on Kolovrat Ridge south of the Isonzo had been smashed as far west as the Luico pass, and to the east to Hill 1114 principally through the combat activities of the Württemberg Mountain Battalion. This action permitted the Alpine Corps and the units of the 12th Division north of Luico to continue their advance.

Chapter 12

THE THIRD DAY OF THE TOLMEIN OFFENSIVE

I: The Assault on Mount Cragonza

With the units of the Württemberg Mountain Battalion located near Luico, I hurried back to our road block north of Polava where I reorganized my detachment, consisting of seven companies, and divided the captured pack animals among the various companies. Without taking time for rest we began climbing in the direction—Jevszek-Cragonza, for we felt that an immediate attack would catch the enemy unprepared.

In spite of the tremendous exertion and privations of the past few days, we soon gained altitude in the steep, trackless terrain that went up partly over long meadows and along impenetrable thorn hedges, partly up through stony draws. Once again I had to demand superhuman efforts from my exhausted troops, for the offensive had to continue.

The ascent became more difficult the higher we climbed. Deep gullies and thorn bushes forced us to make detours which usually resulted in a loss of height and in increased expenditure of energy. We climbed for hours, twilight fell and then darkness. The troops were totally exhausted. Did I relinquish my objective? No, Jevszek had to be reached and once there I knew I would still find enough brave men to storm Mount Cragonza.

The great disk of the moon shone brightly on the roof-steep slope, silvered the grassy surfaces and bushes, and threw long black shadows on the end behind the groups of trees. The point climbed slowly and carefully and finally found a narrow footpath. The detachment followed at fifty-yard intervals. We halted occasionally and listened into the night for any telltale noise.

Once again we halted in the shadow of a haystack close to the narrow path. A densely overgrown ravine lay in front of us and dark shades were uncanny. Our path went through it. We listened intently and we heard the hum of voices, commands, and the noise of marching troops coming from the far side. The enemy was not coming any nearer, but was moving laterally on the far side of the ravine. We thought he might be in position over there and the narrow footpath was our only means of approach. The situation was none too inviting. I figured that Jevszek and Mount Cragonza were up ahead and to our right. The point worked its way up the steep slope in the shadow of the long rows of bushes. Before us lay a large grassy field, brightly illuminated by the moon and surrounded in a semicircle by tall trees. Was it an illusion? Were those obstacles at the edge of the wood? We crept forward with utmost care and found that we had not been deceived. And then we heard Italian voices in the woods to our front. Unfortunately we were unable to 'etermine whether or not the enemy had already occupied prepared positions. To make certain I sent out several officer reconnaissance detachments. Meanwhile the detachment closed up and rested. Soon the report arrived that the enemy was preparing to occupy the positions in front of us, and that the obstacles in front of the position were very high.

To attack that fortified position uphill across the brightly illuminated surface was most daring even for completely fresh troops. With the exhausted mountain soldiers, who had accomplished real wonders since the start of the offensive, such an attack, even in the next few hours, was impossible. Moreover it was still debatable whether a breakthrough at this place in the early part of the night would be useful and could be sufficiently exploited. I gave up the idea, decided on several hours of rest, and undertook a very thorough reconnaissance of terrain and enemy.

Without a sound I moved the detachment into a broad draw which offered protection against fire from above, and was some three hundred yards from the enemy position, and made provisions to rest until midnight. The 4th and 2d Companies secured the bivouac by posting sentries in a semicircle. Since the captured pack animals made themselves unpleasantly conspicuous by whinnying, they were tethered considerably lower down. During the movement to the resting place, very violent combat [the 1st Battalion of the Bavarian Infantry Life Guards had run into an Italian position. Translator.] broke out in the valley near Polava, which indicated that hostile forces were still in the valley.

Other officer-scout squads were sent to reconnoiter favorable avenues of approach into the hostile position, the strength of the obstacles, their depth, possible guns, the type of garrison, and the location of the village of Jevszek. They had to report back by midnight at the latest.

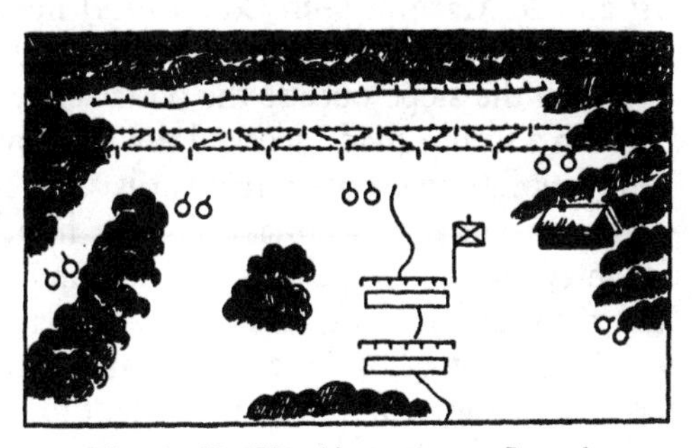

Sketch 47: The bivouac near Jevszek.

I went to sleep just above the detachment in a captured Italian sleeping bag which my hard-working orderly Reiber had discovered on one of the captured mules. In spite of exhaustion, my nervous tension made sleep impossible. As early as 2230 the following excellent report of Lieutenant Aldinger brought me to my feet:

"Jevszek is half a mile northwest of our bivouac area. The town is strongly fortified and equipped with wire entanglements all round, but as yet does not seem to be occupied by the enemy. On the slope just west of Jevszek, as well as through the south part of Jevszek, Italian troops are working downhill in a southeasterly direction."

My decision was quickly made: "On to Jevszek!" I hoped to reach the town ahead of the Italian garrison intended for it. It took only a few minutes to break camp, pull in our security elements and get the companies ready to march. Meanwhile the moon set; the night was dark with the stars furnishing a dim light.

The detachment climbed noiselessly toward Jevszek on the path reconnoitered by Lieutenant Aldinger. The leaders were briefly instructed as to the situation. The 4th Company and 3d Machine-Gun Company formed the advance guard and the other five companies followed at short intervals. First we crossed a narrow bit of wood, then we climbed steeply up the mountain through forest glade. The point soon reached some six-foot obstacles. Lieutenant Aldinger announced that we were only three hundred yards from Jevszek. We halted and listened intently for minutes into the darkness. Nothing moved in our immediate vicinity, but some one hundred yards upslope we heard the tread of descending Italian infantry.

Lieutenant Aldinger slipped through the narrow passage in the wire into the position behind it and found it empty. The point followed. I then moved the entire advance guard in and disposed it in a semicircle inside the hostile installation. Scout squads were sent out to explore the immediate terrain and to reconnoiter against the enemy on the slope and the village of Jevszek.

At the same time the bulk of the detachment (2d and 3d Companies, 1st and 2d Machine-Gun Companies) moved through the obstacle and into the position. I left the signal company and the pack-animal detachment on the slope outside the obstacle.

With a scout squad I worked my way toward the enemy up the slope. Our visibility was only a few yards. The slope in front of us appeared as an uncanny, black mass. A bare hundred yards away the Italian infantry was moving along, probably in column of files, from the right down toward Jevszek on the left. We crept closer. Suddenly we were challenged by a hostile sentry. That was the set-up. The enemy was in position, with a column in movement to his rear.

We crept back and turned to the left toward Jevszek. As we reached the first houses, a scout squad returned with the report that the north part of Jevszek was free of the enemy but that Italian infantry was marching through the south part of the village. I decided to move into

Jevszek with the purpose of capturing the infantry in the southern part.

A few minutes later the detachment moved slowly forward toward the town. The foremost units had just reached the first houses when the dogs in several farms began to bark. Shortly afterwards the enemy opened fire from a position on the slope up to the right and about one hundred yards away. The volley luckily struck mainly in the forest to our left. Not finding cover, we flattened ourselves on the ground, our machine guns and carbines ready for fire, but remained absolutely quiet. The opening of fire from our side was out of the question unless the enemy attacked. If he did not attack, which I considered possible, then he would soon cease firing thinking he had made a mistake.

During the firing units of the main force moved into Jevszek under cover of the unoccupied positions east of the town. The enemy's fire died after a few minutes and my detachment was soon in the town. Happily, no losses resulted from the hostile fire attack.

I occupied the north part of the town in a semicircle, avoiding any further encounter with the enemy on the slope just to the northwest of Jevszek. Midnight was long past. Those not on sentry duty or in position rested on their arms in the houses which were still occupied by Slovene families. We all knew that we were only a hand-grenade's throw from a strongly-occupied Italian position, and that we would be engaged in hand to hand fighting at any moment should the enemy feel his way into the town.

Since the firing the march of hostile forces on the slope northwest of Jevszek and through the south part of the village had ceased. Besides, the enemy had fired on us only from the slope to the northwest and not a single shot came from the southern part of the town. This raised the suspicion that there might be a gap in the enemy's position which we thought was continuous as far as Polava. By the flickering light of an open fire in one of the houses I studied the map thoroughly. We were about a mile north of Polava in the northern part of Jevszek at about twenty-eight feet elevation. Mount Cragonza was six hundred yards to the west and nine hundred feet higher up. Since Jevszek was fortified on the east side, and the enemy was in position on the slopes northwest of Jevszek and southeast as far as Polava, we were dealing with an Italian rearward position prepared long ago with the aim of blocking a penetration through the Luico pass. The hostile movements detected by us during the night led us to believe that the Italians were making every effort to occupy this position. Judging by the type of fortification, Jevszek itself doubtlessly belonged to this position. For some reason the garrison destined for Jevszek had not arrived. We could expect its arrival at any time. Should we wait? I felt that the god of War was once more offering his hand to our courageous mountain troops. I believed that our seizure of Jevszek had given us a portion of those hostile positions which were to block our way and that of the Alpine Corps toward Mount Cragonza, the Mrzli and the Matajur.

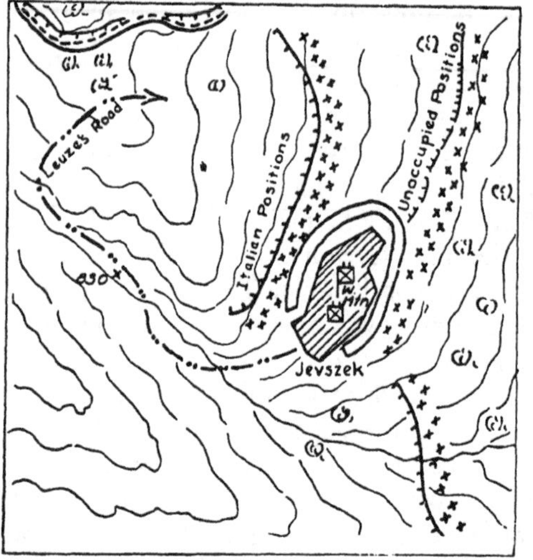

Sketch 48: The situation at Jevszek before daybreak, October 26, 1917.

After these considerations, I told Lieutenant Leuze to go and see if the southwest part of Jevszek was free of the enemy, in which case he was to extend his reconnaissance to the ridge six hundred yards northwest of Jevszek and in the rear of the Italian forces in position just northwest of the village. He was to return in two hours. Lieutenant Leuze declined assistance and moved off alone.

The exhausted detachment was given another rest period. The bulk of the unit sat, within a few yards of the enemy, before the hearthfires in solidly built houses, consuming coffee and dried fruit, which were offered to us by the very friendly Slovenes. An occasional shot was heard outside, followed by an Italian hand-grenade burst. The enemy evidently lacked any desire for a reconnaissance thrust toward Jevszek. We did not fire a shot. Complete darkness enveloped the German and Italian forces so closely opposed to each other.

Toward 0430 Lieutenant Leuze returned from his reconnaissance with an Italian captive, and reported: "The southwest end of Jevszek is free of the enemy, the path to the height six hundred yards northwest of Jevszek has been reconnoitered; I captured this Italian on that hill, but otherwise failed to meet the enemy." He had splendidly fulfilled his mission.

Leuze's report led me to decide on an immediate occupation of the hill six hundred yards northwest of Jevszek with four companies, leaving the remainder of the detachment in Jevszek as support. I planned to attack the enemy northwest of Jevszek at dawn.

This was not an easy decision. If the enemy used his commanding positions on Mount Cragonza to pour fire into ours, then we would have been faced with a fight on two fronts.

It was still pitch dark when, at 0500 the 2d and 4th Rifle Companies and the 1st and 2d Machine-Gun Companies silently left Jevszek and took the path reconnoitered by Leuze. Lieutenant Leuze led at the head of the long file column. I left the 3d Company and 3d Machine-Gun Company under the tested Lieutenant Grau as support in Jevszek with the task of pinning down the garrison of the position northwest of Jevszek by fire as soon as we attacked. Their subsidiary mission was to protect us from an attack from the east.

I gave these orders while the detachment was moving out of the village. When I later joined the column of the 2d Machine-Gun Company, it was getting light on Mount Cragonza. In the mountains the change from night to day takes little time. I had the uncomfortable feeling of being half an hour too late. In front of me I saw my companies climbing Hill 830 in the usual file column among jumbled boulders in the bare hollow. The uppermost crags of Mount Cragonza were already bathed in bright light. I studied them with the glass and became alarmed! Enemy positions were located a few hundred yards above and to the left of my detachment. They were occupied and I could even see the helmets of the garrison. If the enemy opened fire, the hollow in which the detachment was located at the moment offered little cover and heavy losses were unavoidable. At this moment the responsibility for the lives of my officers and men weighed very, very heavily on me; I had to extricate them from the danger of which they were oblivious.

I gathered as much of the 2d Machine-Gun Company as was possible and emplaced it on the left with instructions to cover and pin down the enemy on the slopes up on the left with fire as soon as he began to shoot. Then I hurried forward with the runners and turned the heads of the various companies to the right toward the height overgrown with single small bushes six hundred yards northwest of Jevszek. It was high time; dusk was giving way to daylight.

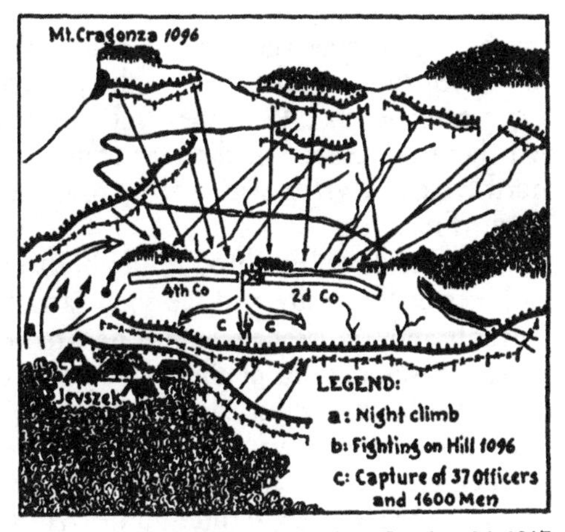

Sketch 49: The fight on Hill 1096, October 26, 1917. View from the east.

As the last units of the companies left the hollow, the enemy on the Cragonza covered the detachment with heavy rapid fire. We were still on the slope facing the enemy and the fire came from a commanding position. There was no protective cover. Only low thorn bushes in various places offered the possibility of at least withdrawing from the enemy's sight. Under the quickly initiated fire support of the 2d Machine-Gun Company the various platoons split, gained the height six hundred yards northwest of Jevszek, and took up the fight from there.

But we were not equal to the powerful superiority of the hostile fire directed against us from a semicircular position on the heights in the northwest, west, and southwest. By creeping laterally and making short rushes, the men of the 2d and 4th Companies tried to separate and reduce the effect of the hostile fire. Losses mounted. Among others, the fine leader of the 2d Company, Lieutenant Ludwig, was seriously wounded.

Meanwhile to our rear the battle in Jevszek had flared up. According to orders, the 3d Company and the 3d Machine-Gun Company under Lieutenant Grau had taken the enemy northwest of Jevszek under fire, pinned him to his positions and prevented him from attacking the rear of the other companies.

With a few combat orderlies I reached the hill six hundred yards northwest of Jevszek where I found cover against aimed fire in a small group of bushes. The machine guns rattled away on all sides. I no longer had as much as a squad in reserve. Everyone was engaged in the violent fire fight and was shooting as rapidly as possible. I had to make a quick decision or lose my unit. Through my combat orderlies I had three light machine-gun squads gathered together from the front line of the 2d and 4th Companies and brought to the protecting slope sixty yards east of my command post. With these men I formed several assault teams and led them downslope in the rear of the enemy located in position just northwest of Jevszek with a front to the east, who was being subjected to our fire.

We went downhill through the bushes with our machine guns and carbines at the ready and we soon saw the hostile position below us. It was heavily garrisoned, helmet next to helmet. From above we looked down on the bottom of the trench. The enemy had no cover against our fire. Above us whistled the fire intended for the mountain troops on the hill six hundred yards northwest of Jevszek. Down below and near Jevszek we saw the 3d Company and the 3d Machine-Gun Company firing on the Italians just three hundred feet below us. The enemy did not suspect what threatened him.

The assault squads made ready and we shouted down to the hostile garrison and told them to surrender. Frightened, the Italian soldiers stared up at us to the rear. Their rifles fell from their hands. They knew they were lost and gave the sign of surrender. My assault squads did not fire a single shot. Not only did the garrison of the positions between us and Jevszek, about three companies strong surrender; but, to our great surprise, the hostile trench garrison as far north as the Matajur road also laid down its arms. It had been completely confused by the fierce noise of battle in its rear and by the appearance of the weak assault squads on the northeast slope of the hill six hundred yards northwest of Jevszek. The fire fight between the Italian garrisons on Mount Cragonza and the greater part of the Rommel detachment probably meant to the enemy that the Germans were attacking in the direction from Cragonza and had already occupied their commanding heights.

An Italian regiment of 37 officers and 1000 men surrendered in the hollow seven hundred yards north of Jevszek. It marched up with full equipment and armament, and I had trouble finding enough men to carry out the disarmament. Meanwhile the battle raged on some three hundred feet above with undiminished violence.

The Italian garrison on Mount Cragonza knew nothing of the activities in the vicinity of Jevszek and it continued to attack my front line. But our rear had been cleared.

The companies released near Jevszek moved up and soon launched a frontal attack against Cragonza. It was a tough fight. The enemy clung tenaciously to his strong, commanding positions, against which our fire had but little effect. The mountain troops moved across the bare, steep slopes through a hail of lead and closed with the enemy.

Since I had no more forces to commit, I advanced with the 2d Company (in the middle). It was led by Lieutenant Aldinger in place of the severely wounded Lieutenant Ludwig. We reached the lower loop of the Matajur road and found fourteen teamless Italian field guns and twenty-five ammunition wagons. They might have been the artillery groups from Avsa and Perati. We had no time to waste there. Flanking machine-gun fire from the north struck amongst us. We rushed on. Shortly afterward the 2d Company lost its new commander when Lieutenant Aldinger was severely wounded by three bullets. For a while on the Matajur road I myself provided a target for an Italian machine gunner.

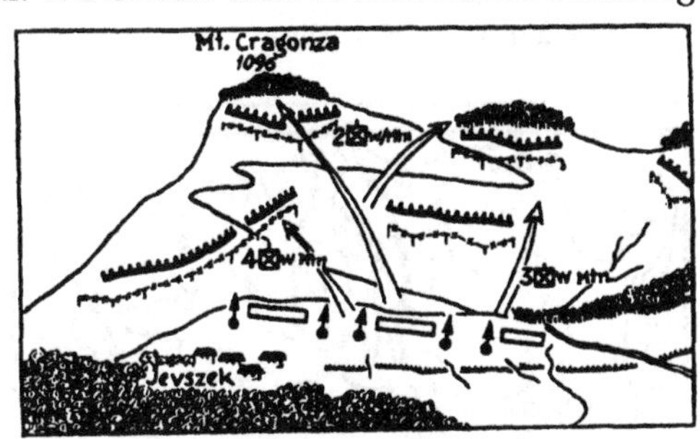

Sketch 50: The attack against Mount Cragonza. View from the northeast.

There was no cover against his fire. I escaped the well-laid cone of fire by running upslope around a bend of the road about seventy yards away.

The losses only increased the fury of the mountain troops. Trench after trench and machine-gun nest after nest were taken. The hard job was completed at 0175. The valiant 2d Company, now led by Technical Sergeant Hügel, had captured the peak of Mount Cragonza. Thus the fate of the hostile forces on the northeast and east slopes of the Mrzli peak was merely a question of time.

I could only surmise as to our neighbor's progress. To judge by the sound which had been increasing continually on our right since dawn, I imagined that units of the 12th Division and the Alpine Corps were attacking Mrzli peak from the northeast and east, perhaps they were also climbing the Cragonza from Avsa along the Matajur road.

I debated the advisability of awaiting their arrival or of reorganizing my detachment which had become badly mixed. My men had really earned a rest, but I had to consider the possibility of being counterattacked by Italian reserves who were still on my right flank.

I considered it best to anticipate hostile countermeasures by continuing the attack without delay against the ridge leading to Mrzli pass with all available forces (half a company).

Observations: In the nocturnal ascent toward Jevszek the Italian troops betrayed themselves by shouting and noisy marching. Thus they put us on the right road and we succeeded in avoiding an undesired encounter.

While the exhausted troops rested, the officers were untiringly active in determining precise information regarding the enemy and the terrain. Even after midnight they continued reconnoitering from Jevszek. Thus they created the basis for the successful penetration northwest of Jevszek and for the taking of Mount Cragonza.

In that night of October 25 I knew very little about my neighbors. I did not know where they were, what they were doing, or what they planned to do. There was also no contact with the outposts. But it was clear to me that this had to be accepted in order to get the attack going again on October 26.

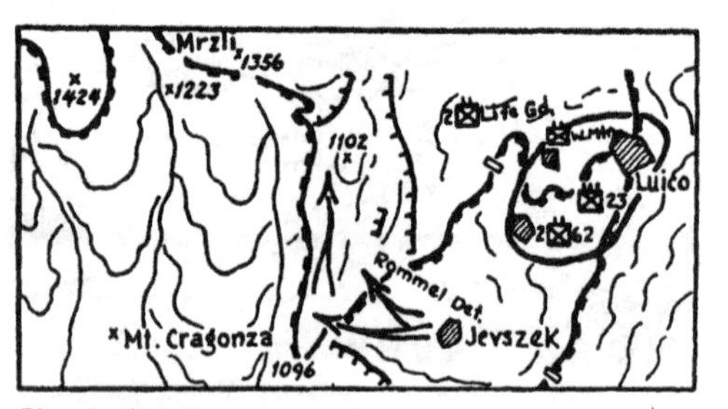

Sketch 51: Storming Hill 1096, October 26, 1917.

Even the very desperate situation of the mountain troops at daybreak under fire and without cover between the hostile positions finally turned in our favor. A few squads of brave men brought about this change. The attacking power of the Württemberg mountain troops was shown particularly in the frontal attack against the Italians in commanding and excellent positions on Mount Cragonza. It did not fail in the 2d Company even after all its officers became casualties.

At 0715 on October 26, 1917, at the time of the taking of Mount Cragonza the order of battle was:

Kraus Group: The Stol (1668 m.) fell in the night of October 25-26 at 0300, to the 2d Battalion of the 1st Imperial Rifle Regiment which reached Bergogna at 0600. The 1st and 3d Battalions of that regiment and the 43d Brigade followed and reached Bergogna at 0800.

Stein Group: The 12th Division had the 63d Infantry Regiment at the border in the Natisone valley as on the day before; the 2d Battalion of the 62d Infantry Regiment and the 23d Infantry Regiment closed on the outposts of the 2d Life Guards Battalion at Avsa and were ready to move.

Alpine Corps: The Württemberg Mountain Battalion broke into the hostile position: Mrzli peak—Jevszek—Polava at Jevszek, opened this position eleven hundred yards to the northwest and took Mount Cragonza at 0715. The rest of the Württemberg Mountain Battalion marched from Luico through Avsa to Mount Cragonza. The 2d and 3d Battalions of the Life Guards prepared to march and later joined the 1st and 3d Companies of the Württemberg Mountain Battalion for the advance on Mount Cragonza. The 1st Battalion of the Life Guards was on outpost duty at Polava. The 2d Jäger Regiment (less the 10th Company) moved from Ravna to Luico. The 1st Jäger Regiment and 10th Jäger Company prepared to march from Hill 1114, where they had spent the night. The 200th Division: The 3d Jäger Regiment advanced through Drenchia to Trusgne, reaching the latter at 0800. The 4th and 5th Jäger Regiments moved to Ravna at 0430 from Hill 1114, where they had spend the night. They remained in Ravna until 0800.

Scotti Group: The 8th Grenadier Regiment used its 1st Battalion to take La Kalva at 0500 and then attacked Mount Hum with all three battalions.

The net result was: The Italian position (north slope of the Matajur—Mrzli peak—Jevszek—Polava—St. Martino) was, as was the Kolovrat Ridge position, smashed near Jevszek in the early hours of the morning by the leading units of the Württemberg Mountain Battalion, and consequently Mount Cragonza, the key to all Italian positions on the Mrzli peak and Matajur massif, was taken.

II: The Capture of Hill 1192 and the Mrzli Peak (1356) and the Attack on Mount Matajur

In spite of the exhaustion following the seizure of Mount Cragonza, I was unable to give my men a well-earned rest on the summit. The splendid Technical Sergeant Hügel took up his new job with his characteristic energy, exploited his limited forces to a maximum and, without waiting for support, attacked along the ridge rising toward Hill 1192 (3840 feet) and Mrzli peak in order to gain additional ground.

I sent orders by runners to the detachment to follow quickly over Mount Cragonza and to take the Matajur road in the direction of Mrzli peak. Then I joined the 2d Company. A hundred yards farther on we ran into the enemy who was dug in on a wooded knoll on the ridge. On the east slope to our right the noise of combat increased considerably. Apparently rearward units of the Rommel detachment, climbing from Jevszek toward Cragonza, were being fired on or attacked. But it might have been units of the Alpine Corps attempting to climb the Matajur Road from Luico to Mount Cragonza.

Technical Sergeant Hügel was a past master at holding the enemy, who was superior in numbers and weapons, frontally and simultaneously attacking him in flank and rear with assault squads. These movements were accomplished in a few minutes and led to a repulse of the enemy, causing him to retire to the northeast, downhill toward Luico.

Since we attacked readily whenever we met the enemy, contact with our rear was soon broken. A report reached us that the detachment was being delayed by strong machine-gun fire from Italian positions northeast of Cragonza and was almost a mile behind me. I decided not to halt the 2d Company but to continue the attack against Mrzli peak until we encountered a strong enemy.

By 0830 the 2d Company, having dwindled to a platoon with two light machine guns, captured Hill 1192 a mile and a half west of Avsa. The enemy prevented a further advance. He was in considerable strength half a mile northeast of Mrzli peak (1356) (3480) and plastered our newly won hilltop with heavy machine-gun fire. Lively fighting was in progress down the slope to the right, and also to the right rear in the direction of Jevszek. Alpine Corps units were attacking.

A minimum force required to attack the enemy on the southeast slope of the Mrzli was estimated at two rifle companies and one machine-gun company. In order to assemble these forces quickly, I hurried to the rear down the Matajur road. Hügel had orders to hold Hill 1192. Even after searching far and wide I found no liaison officer of the trailing Rommel detachment. After rounding a curve seven hundred yards south of Hill 1192, I suddenly ran into an Italian detachment which was coming from the direction of Avsa and was crossing the Matajur road. The Bersaglieri grabbed their rifles and fired. A quick leap into the bushes just below the road saved me from the aimed fire. A few adversaries followed me downslope through the bushes. But while they hastened down toward the valley, I was climbing toward Hill 1192. Arriving there, I ordered a fairly-strong scout squad to establish contact with the other units of the Rommel detachment and to give the various company commanders the order to close up Hill 1192 as soon as possible. Note: Meanwhile units of the Alpine Corps and the 12th Division which were in the Perati—Avsa—Luico area, had started to march down the Matajur road in the direction of Mount Cragonza. The 2d Battalion of the 62d Infantry Regiment, marching at the head, encountered enemy in strong positions a mile south of Avsa and attacked him. The units following it (the staff and the Gössler detachment of the Württemberg Mountain Battalion, the 23d Infantry Regiment, and the 2d and 3d Battalions of the Life Guards) succeeded in advancing along the Matajur road in the direction of Mount Cragonza. The 1st Battalion of the Life Guards was still stalled by an Italian blocking position near Polava.

I had to wait until 1000 before I had assembled a force equal to two rifle and one machine-gun company. These groups were composed of all companies of the Rommel detachment. Their approach to Hill 1192 was greatly delayed, because the various units were repeatedly involved in battles with the enemy, who was trying to retreat in a southwesterly direction across the Mount Cragonza—Hill 1192 line.

I felt we were strong enough to engage the Italian garrison on the Mrzli. By means of light signals we asked for artillery fire on the hostile positions on the southeast slope of the Mrzli peak, with the astounding result that German shells were soon striking there. Then the

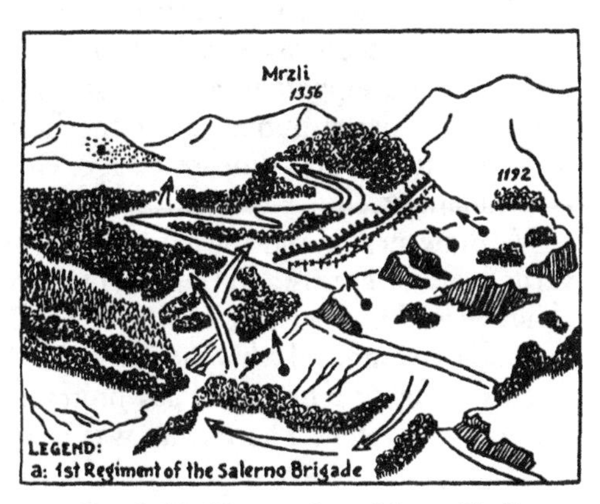

Sketch 52: The attack on Mount Mrzli.

lively fire of the machine-gun company from Hill 1192 pinned the hostile garrison down in their positions while two rifle companies under my leadership came into close combat with the enemy just below the ridge road. We succeeded in turning the hostile west flank. Then we swung in against the flank and rear of the hostile position. But the enemy hastily withdrew when he saw us attacking in this direction and retired to the east slope of the Mrzli. We took a few dozen prisoners. Since I did not intend to follow the enemy retreating on the east or north slopes of the Mrzli, I broke off the engagement, continued my advance down the ridge road toward the south slopes of the Mrzli and brought up the machine-gun company.

Already during our attack we had observed hundreds of Italian soldiers in an extensive bivouac area in the saddle of the Mrzli between its two highest prominences. They were standing about, seemingly irresolute and inactive, and watched our advance as if petrified. They had not expected the Germans from a southerly direction—that is, from the rear. We were only a mile away from this concentration of troops. The Matajur road wound up over the partially-wooded south slope of the Mrzli and, on the way west to the Matajur, passed just under the hostile encampment.

The number of the enemy in the saddle on the Mrzli was continually increasing until the Italians must have had two or three battalions massed there. Since they did not come out fighting, I moved nearer along the road, waving a handkerchief, with my detachment echeloned in great depth. The three days of the offensive had indicated how we should deal with the new enemy. We approached to within eleven hundred yards and nothing happened. He had no intention of fighting although his position was far from hopeless! Had he committed all his forces, he would have crushed my weak detachment and regained Mount Cragonza. Or he could have retired to the Matajur massif almost unseen under the fire support of a few machine guns. Nothing like that happened. In a dense human mass the hostile formation stood there as though petrified and did not budge. Our waving with handkerchiefs went unanswered.

We drew nearer and moved into a dense high forest seven hundred yards from the enemy and thus out of his line of sight, for he was located about three hundred feet up the slope. The road bent sharply to the east and we wondered what the enemy up there would do. Perhaps he had decided to fight? If he rushed downhill we would have had a man to man battle in the forest. The enemy was fresh, had tremendous numerical superiority, and moreover enjoyed the advantage of being able to fight downhill. Under these conditions I considered it a vital necessity to reach the edge of the wood below the hostile camp. But my mountain troopers with the heavy machine guns on their backs were so exhausted that I did not expect them to make the steep climb through dense underbrush.

Therefore I allowed the detachment to continue marching along the road while Lieutenant Streicher, Dr. Lenz, a few mountain soldiers and I climbed on a broad front, about a hundred yard interval between men, and took the shortest route through the forest toward the enemy. Lieutenant Streicher surprised a hostile machine-gun crew and took it prisoner. We reached the edge of the forest unhindered. We were still three hundred yards from the enemy above the Matajur road; it was a huge mass of men. Much shouting and gesticulating was going on. They all had weapons in their hands. Up front there seemed to be a group of officers. The leading elements were not expected for some time and I estimated them to be at the hairpin turn seven hundred yards to the east.

With the feeling of being forced to act before the adversary decided to do something, I left the edge of the forest and, walking steadily forward, demanded, by calling and waving my handkerchief, that the enemy surrender and lay down his weapons. The mass of men stared at me and did not move. I was about a hundred yards from the edge of the woods, and a retreat under enemy fire was impossible. I had the impression that I must not stand still or we were lost.

I came to within 150 yards of the enemy! Suddenly the mass began to move and, in the ensuing panic, swept its resisting officers along downhill. Most of the soldiers threw their weapons away and hundreds hurried to me. In an instant I was surrounded and hoisted on Italian shoulders. *"Evviva Germania!"* sounded from a thousand throats. An Italian officer who hesitated to surrender was shot down by his own troops. For the Italians on Mrzli peak the war was over. They shouted with joy.

Now the head of my mountain troops came up along the road from the forest. They moved forward with their habitual easy but powerful mountaineer stride in spite of the hot sun and their heavy loads. Through an Italian who spoke German I ordered the prisoners to line up facing east and below the Matajur road. There were 1500 men of the 1st Regiment of the Salerno Brigade. I did not let my own detachment halt at all, but I did call one officer and three men out of the column. Two mountain riflemen were assigned to move the Italian regiment across Mount Cragonza to Luico; and the disarming and removal of the 43 Italian officers, separated from their men, was entrusted to Sergeant Göppinger. The Italian officers became pugnacious after seeing the weak

Rommel detachment and they tried to reestablish control over their men. But now it was too late. Göppinger performed his duty conscientiously.

While the disarmed regiment moved down toward the valley, the Rommel detachment moved past just below the Italian camp ground. Some captured Italians had told me shortly before that the 2d Regiment of the Salerno Brigade was on the slopes of the Matajur; it was a very famous Italian regiment which had been repeatedly praised by Cadorna in his orders of the day because of outstanding achievements before the enemy. They assured me that this regiment would certainly fire on us and that we would have to be careful.

Their assumption was correct. The head of the Rommel detachment no sooner reached the west slope of the Mrzli than strong machine-gun fire opened up from Hills 1467 (4842) and 1424 (4700). The hostile machine-gun fire was excellently adjusted on the road and soon swept it clear. Dense bushes below the road protected us from aimed fire. My men were soon under control and I continued the march, not below the Matajur road in the direction of Hill 1467 but in a sharp turn to the southwest. I wanted to cross Hill 1223 (4035) at the double and head toward the hairpin turn in the Matajur road just south of Hill 1424. Once there, then the 2d Regiment of the Salerno Brigade could scarcely escape and would be in a position similar to that of the 1st Regiment a half hour before. The only difference would be that a withdrawal to the south across the bare slopes of the Matajur would be prevented by our fire, whereas on Mrzli peak a covered retreat through the wooded zone had remained open to the Italians.

In order to deceive the enemy, I ordered a few machine guns to fire from the west slopes of the Mrzli. With the rest of the detachment I reached the turn of the road seven hundred yards south of Hill 1424 without undergoing hostile fire, for the enemy was unable to observe our movement through thick clumps of bushes. I prepared a surprise attack on the garrison of Hill 1424, which was still firing on the rearward units of the Rommel detachment and on our machine guns on the Mrzli. The success of the Mrzli had caused us to forget all our efforts, our fatigue, our sore feet and our shoulders chafed by heavy burdens.

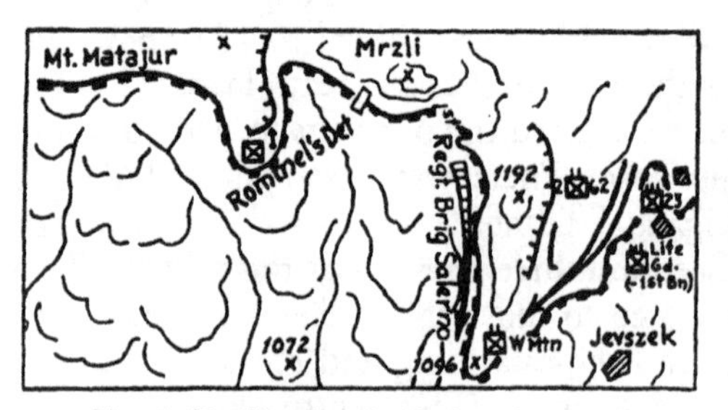

Sketch 53: The situation before the attack on Mount Matajur.

While I was expeditiously carrying out the preparations for the attack, ordering the machine-gun platoons in position, and organizing assault squads, the order came from the rear: "Württemberg Mountain Battalion withdraws." (Major Sprösser had reached Mount Cragonza. The great number of prisoners of the Rommel detachment (over 3200 men) had reached him and given the impression that the hostile resistance on the Matajur massif was already broken.) The battalion order to withdraw resulted in all units of the Rommel detachment marching back to Mount Cragonza, except for the hundred riflemen and six heavy machine-gun crews who remained with me. I debated breaking off the engagement and returning to Mount Cragonza.

No! The battalion order was given without knowledge of the situation on the south slopes of the Matajur. Unfinished business remained. To be sure, I did not figure on further reinforcements in the near future. But the terrain favored the plan of attack greatly and—every Württemberg Mountain trooper was in my opinion the equal of twenty Italians. We ventured to attack in spite of our ridiculously small numbers.

Over on Hills 1424 and 1467 the defender was facing east among large rocks and he dove for cover when our unexpected machine-gun fire hit him from the south. The heavy fragmentation up there in the rocks considerably increased the effect of each shot. The hostile reaction was slight. Our machine guns had been emplaced in dense, high bushes, so that the enemy had trouble locating them.

I observed the splendid effect of our fire with the glass. When the first Italians tried to retire to the north slope of Hill 1424, I advanced my riflemen astride the Matajur road and on the west slope of Hill 1424. We advanced rapidly thanks to the strong fire support from the heavy machine guns. Over on the right the enemy completely vacated his positions on the east slope of Hill 1424 and his fire died out.

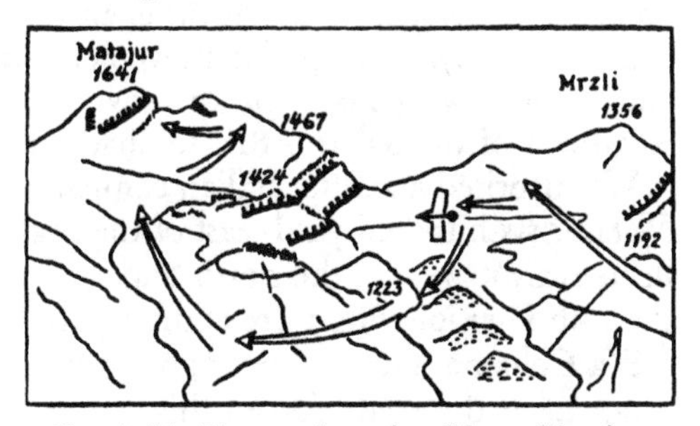

Sketch 54: The attack against Mount Matajur.

We kept attacking. The heavy machine guns were moved up in echelon. From Hill 1467 a hostile battalion tried to move off to the southwest by way of Scrilo. But the fire of one of our machine guns, delivered at sixty yards from the head of the column, forced the battalion to halt. A few minutes later, waving handkerchiefs, we approached the rocky hill six hundred yards south of Hill 1467. The enemy had ceased firing. Two heavy machine guns in our rear covered our advance. An unnatural silence prevailed. Now and then we saw an Italian slipping down through the rocks. The road itself wound among the rocks and restricted our view of the terrain to a few yards. As we swung around a sharp bend, the view to the left opened up again. Before us—scarcely three hundred yards away—stood the 2d Regiment of the Salerno Brigade. It was assembling and laying down its arms. Deeply moved, the regimental commander sat at the roadside, surrounded by his officers and wept with rage and shame over the insubordination of the soldiers of his once-proud regiment. Quickly, before the Italians saw my small numbers, I separated the 35 officers from the 1200 men so far assembled, and I sent the latter down the Matajur road at the double, toward Luico. The captured colonel fumed with rage when he saw that we were only a handful of German soldiers.

Without stopping I continued the attack against the summit of the Matajur. The latter was still a mile away and seven hundred feet above us and we could see the garrison in position on the rocky summit. It apparently did not intend to follow the example of its comrades on the south slope of the Matajur who had surrendered and were marching away. Lieutenant Leuze used his few machine guns to give fire support for the attack which we attempted on the shortest route from the south. But the hostile defensive fire was very heavy there and the avenues of approach were so disadvantageous that I preferred to turn to the east on the arched slope, unseen by the enemy, and attack the summit position from Hill 1467. During this movement small squads of Italians, with and without weapons, kept on moving toward the spot where the 2d Regiment of the Salerno Brigade had laid down its arms.

We surprised an entire Italian company on the sharp east ridge of the Matajur six hundred yards east of the peak. In total ignorance of events in its rear, it was on the north and was engaged with scout squads of the 12th Division who were climbing toward the Matajur from Mount Della Colonna. Our sudden appearance on the slope in the rear with weapons at the ready forced this enemy to surrender at once without resistance.

While Lieutenant Leuze fired on the garrison of the summit with a few machine guns from a southeasterly direction, I climbed with the other units of my small group in a westerly direction along the ridge and toward the summit. On a rocky knoll a quarter of a mile east of the peak, other heavy machine guns went into position as fire support for the assault team disposed on the south slope. But before we opened fire, the garrison of the summit gave the sign of surrender. One hundred and twenty more men waited patiently until we took them prisoner at the ruined building (border guardhouse) on the summit of the Matajur (1641) and (5415). A scout squad of the 23d Infantry Regiment, consisting of a sergeant and six men, met us during its climb from the north.

At 1140 on October 26, 1917, three green and one white flare announced that the Matajur massif had fallen. I ordered a one-hour rest on the summit. It was well deserved.

Round about we saw the mighty mountain world in radiant sunshine. Our view reached far: In the northwest the Stol lay six miles away and was being attacked by the Flitsch Group. In the west we saw Mount Mia (1228) far below us. We could not see into the Natisone valley, though it lay only two miles away and forty-seven hundred feet below us. In the southwest were the fertile fields around Udine, Cadorna's headquarters. In the south the Adriatic glittered. In the southeast and east were the mountains so well known to us: Cragonza, Mount San Martino, Mount Hum, Kuk, Summit 1114.

That war was still round about us was indicated by the prisoners sitting amongst us, by weak artillery fire, and by an air battle, in which an Italian machine plunged burning into the depths. Nothing was to be seen of our neighbors. I dictated the combat report which Major Sprösser demanded every day to Lieutenant Streicher.

Observations: The capture of Mount Matajur occurred fifty-two hours after the start of the offensive near Tolmein. My mountain troopers were in the thick of battle almost uninterruptedly during these hours and formed the spearhead of the attack by the Alpine Corps. Here—carrying heavy machine guns on their shoulders—they surmounted elevation differences of eight thousand feet uphill and three thousand downhill, and traversed a distance of twelve miles in an air line through unique, hostile mountain fortifications.

In twenty-eight hours five successive and fresh Italian regiments were defeated by the weak Rommel detachment. The number of captives and trophies amounted to: 150 officers, 9000 men, and 81 guns. Not included in these figures were the enemy units which, after they had been cut off on the Kuk, around Luico, in the positions on the east and north slopes of the Mrzli peak, and on the north slopes of Mount Matajur, voluntarily laid down their arms and joined the columns of prisoners moving toward Tolmein.

Most incomprehensible of all was the behavior of the 1st Regiment of the Salerno Brigade on the Mrzli. Perplexity and inactivity have frequently led to catastrophies. The councils of the mass undermined the authority of the leaders. Even a single machine gun, operated by an officer could have saved the situation, or at least would have assured the honorable defeat of the regiment. And if the officers of this regiment had led their 1500 men against the Rommel detachment, then Mount Matajur would scarcely have fallen on October 26.

In the battles from October 24 to 26, 1917, various Italian regiments regarded their situation as hopeless and gave up fighting prematurely when they saw themselves attacked on the flank or rear. The Italian commanders lacked resolution. They were not accustomed to our supple offensive tactics, and besides, they did not have their men well enough in hand. Moreover, the war with Germany was unpopular. Many Italian soldiers had earned their living in Germany before the war and found a second home there. The attitude of the simple soldier toward Germany was clearly displayed in his *"Evviva Germania!"* on the Mrzli.

A few weeks later the mountain soldiers had Italian troops opposing them in the Grappa region, who fought splendidly and were men in every particular, and the successes of the Tolmein offensive were not repeated.

The evaluation of the successes of the Württemberg Mountain troops in the first days of the Great Battle is evident in the orders of the day of the German Alpine Corps (General von Tutschek) of November 3, 1917, which states among other things: "The capture of the Kolovrat Ridge caused the collapse of the whole structure of hostile resistance. The Württemberg Mountain Battalion under its resolute leader, Major Sprösser, and his courageous officers was the main one active here. The capture of the Kuk, the possession of Luico, and the penetration of the Matajur position by the Rommel detachment initiated the irresistible pursuit on a large scale."

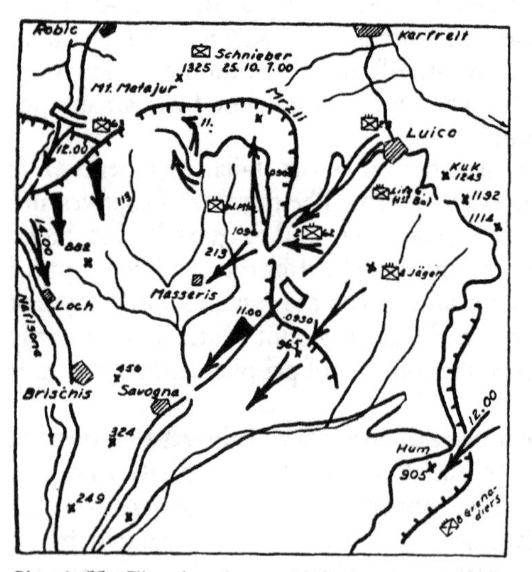

Sketch 55: The situation at 1200, October 26, 1917.

The losses of the Rommel detachment in the three days of attack were happily low: 6 dead, including 1 officer; 30 wounded, including 1 officer.

On October 26, 1917, at midnight the order of the Flitsch-Tolmein battle was:

Kraus Group: Forward units meeting in Bergogna. Hostile attack on the Passo di Tanamea was repulsed.

Stein Group: In the 12th Division sector an attack by the 62d and 63d Infantry Regiments was under way in the Natisone valley from the border across Stupizze toward Loch. The latter was reached around 1400. From the north no forces were in position to attack the Italian position on the Matajur-Mrzli line. The 23d Infantry Regiment marched over Cragonza toward Matajur reaching the Cragonza around noon. In the Alpine Corps the Rommel detachment of the Württemberg Mountain Battalion took the Mrzli and the Matajur. The bulk of the Württemberg Mountain Battalion under Major Sprösser was descending from Mount Cragonza toward Masseris. The 2d and 3d Battalions of the Life Guards followed him; the 1st Battalion of the Life Guards and the 10th Reserve Jäger Battalion started the advance toward Polava at 1000, after the enemy had vacated his positions near Polava. In the 200th Division the 4th Jäger Regiment took Mount San Martino by 0930 and then advanced in the direction of Assida.

Scotti Group: The 8th Grenadier Regiment took Mount Hum during the forenoon. The 1st Imperial and Royal Division continued the attack through Cambresko toward St. Jakob.

The result was: The forces of the 12th Division and the Alpine Corps around Luico advanced in a southwesterly direction only after the Italian positions on Mount Cragonza had been taken and the Salerno Brigade on the Mrzli and the Matajur had been taken by the leading units of the Württemberg Mountain Battalion. Also the attack of the 12th Division in the Natisone Valley northwest of the Matajur massif, which was reached in the night of October 24-25, made gains only after the enemy on the Matajur had been captured.

Chapter 13

PURSUIT ACROSS THE TAGLIAMENTO AND PIAVE RIVERS, OCTOBER 26, 1917—JANUARY 1, 1918

I: Masseris—Campeglio—Torre River—Tagliamento River—Klautana Pass

While we were still on the Matajur, Lieutenant Autenrieth arrived with the battalion order to move to Masseris which lay some twenty-six hundred feet below us. The descent was hard and required the last physical efforts of my dead-tired men. We took the captured officers of the 2d Regiment of the Salerno Brigade along, for they appeared to be intractable and unwilling to accept their new situation and I did not dare to send them to Luico under small guard through terrain covered with thousands of abandoned weapons.

We climbed down along a narrow path and reached the charmingly situated village of Masseris in the early afternoon hours without encountering the enemy. The companies were rapidly distributed among the few farms, the most essential security precautions were taken, efforts were made to re-establish contact with the units of the Württemberg Mountain Battalion, which had marched ahead in the direction of Pechinie, and then the tired troops rested.

I invited the captured officers to a simple supper. No sparkling conversation graced the board and my guests scarcely touched our modest provender. The gentlemen were too shaken by their fate and that of their proud regiment. I understood their plight completely and did not linger at the table.

My detachment was on the road to the Natisone valley well before dawn. The other units of the battalion had moved forward to Cividale and had a considerable head start. While violent fighting was in progress on the heights west of the Natisone, the Rommel detachment moved down the valley toward Cividale without halts or meals. I rode on ahead and, at noon, came up with Gössler's detachment and the Württemberg Mountain Battalion staff near San Quarzo where they were engaged with the enemy, who still held the Purgessino. Lieutenant Streicher and I rode across the battlefield. An occasional burst of Italian machine-gun fire quickened our pace. I met Major Sprösser just east of San Quarzo. My detachment was not committed to action.

Fighting at Purgessimo was over by 1400. After several hours of rest the Rommel detachment moved around midnight into Campeglio from where the remaining units of the Württemberg Mountain Battalion were reconnoitering in the direction of Fadis and Ronchis.

Pursuit was resumed in the early hours of October 28. We pushed west. Rain of cloudburst proportions poured down, soaking us to the skin. For a while the men protected themselves against it with umbrellas which the resourceful fellows had "found" somewhere. Soon, however, higher authority vetoed this new addition to the table of basic allowances. We marched on in the streaming rain without encountering the enemy.

In the afternoon Italian rear guards blocked the road across the swollen stream near Primulacco. As a result of the continuous strong rain the usually shallow brook had become a vicious stream six hundred yards wide. The enemy opposite shot at everything that moved on the east bank.

We moved down into Primulacco, supplied ourselves with dry things in an Italian laundry depot, and went to sleep. The efforts of the last days and night had worn us down greatly. An hour before midnight an order came from Major Sprösser: "Rommel detachment, reinforced by a platoon of mountain artillery, must, during the night or at least before daybreak, force a crossing of the stream." Everybody up! The detachment worked feverishly during the latter half of the night. While the artillery platoon fired several shells on the Italian garrison on the west bank, a footbridge crossing the numerous arms of the stream was constructed of all vehicles that could be moved up. The enemy did not disturb the work very much. He appeared to have withdrawn following the impact of the first shell on the west bank. When day broke, the end of our emergency bridge was an even hundred yards short of reaching the west bank, and the enemy had retired.

Lieutenant Grau was the first to ride through the last and very rapid branch of the stream. Since the supply of requisitioned vehicles was insufficient to reach the west bank, a strong rope was stretched across the last section. The riflemen held onto this while wading through the rapid mountain stream, which undoubtedly would have swept an unaided man away. In crossing, an Italian prisoner carrying a large medical kit on his back was torn from the rope by the strong current and, lying on his back, floated downstream. The man could not swim. Besides, the heavy knapsack dragged him down. I felt sorry for the poor devil. Spurring my horse, I galloped after the Italian, and succeeded in getting near him in the stream. In his deadly fear the Italian seized the stirrup and the good horse brought us both safely to land.

The detachment was across within the next fifteen minutes. We moved through Rizzollo, where the people welcomed us very warmly, and Tavagnaco to Feletto, where we joined the other units of the Battalion, who had crossed on the bridge at Salt. Without hostile contact the battalion moved toward the Tagliamento in the west and reached Pagagua late in the evening. My staff and I drew good quarters. The owners, to be sure, had moved out leaving the house servants behind. We ate and slept.

On October 30 the battalion reached the Tagliamento near Dignano after passing through Cisterna. The local bridge had been destroyed. Strong enemy forces occupied the west bank of the broad and swollen stream and our attempted crossings failed. To the north we found the roads leading through St. Daniel to the bridge at Pietro completely blocked with Italian columns and vehicles of all kinds. Here horse-drawn columns, pack-animal columns, and refugee vehicles were wedged in among truck columns and heavy artillery. Vehicles jammed both sides of the road for miles and were so tangled that none could move forward or backward. Italian soldiers were no longer visible. They had sought safety elsewhere. Horses and pack animals had been wedged in for days, and were so hungry they were eating everything within reach, including blankets, canvas, and leather harness.

A previously planned night advance of the Rommel detachment across the fields to the bridge at Pietro was unfortunately countermanded by higher authority. We regretted missing some excitement and moved to Dignano where we spent the night.

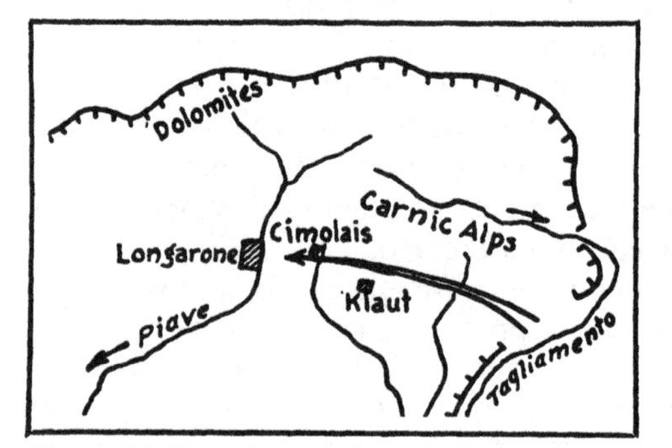

Sketch 56: The advance through the Carnic Alps.

The next day we learned that a unit of the 12th Division had been mentioned in the army communique as having captured Mount Matajur. This matter was soon corrected in higher quarters.

During the succeeding days all attempts to cross the Tagliamento failed. Not until the night of November 2-3, 1917, did the Redl Battalion of the 4th Bosnian Infantry Regiment succeed in getting a foothold on the west bank in the vicinity of Cornino. On November 3d the Württemberg Mountain Battalion was detached from the German Alpine Corps and given the mission of breaking through the Carnic Alps via Meduno-Klaut as advance guard of the 22d Imperial and Royal Infantry Division, and of reaching the upper Piave valley near Longarone as soon as possible in order to force the Italians in the Dolomites to shift their lines of withdrawal to the south.

The Württemberg Mountain Battalion was one of the first to cross the Tagliamento at Cornino. Strong patrols, on captured Italian folding bicycles, advanced to Meduno. Beyond that place, the advance guard of the Württemberg Mountain Battalion succeeded in capturing 20 officers and 300 men near Redona. Then we chased the weak Italian rear guards down a narrow path through the wild and crevassed Klautana Alps toward Klautana Pass. My detachment marched with the main body and Gössler's detachment formed the advance guard. It reached Pecolat in the evening of November 6.

Early on November 7 the Württemberg Mountain Battalion climbed toward Klautana Pass in usual formation. The leading units of the advance guard were fired on from the heights near the pass which had an elevation of forty-nine hundred feet. Machine gun and artillery fire also bothered them on the narrow and winding road between Pecolat and the pass (three thousand feet difference in elevation). Italian fire soon interdicted all movement on the road and through the rocky terrain on either side. The enemy, well dug in, sat high up on the perpendicular rock walls of Mount La Gialina (1634) (5400) and on the northeast ridge of Mount Rosselan (2067) (6822). These two positions were a mile and a half apart and were astride the pass. The position seemed impregnable.

Major Sprösser ordered the Rommel detachment (1st, 2d, and 3d Companies and 1st Machine-Gun Company), which was with the main body, to encircle the enemy on the pass by moving south via Mount Rosselan. Even the climb up the Silisia was much hampered by hostile machine guns and artillery and we were obliged to dash from rock to rock. Finally we reached cover from the hostile fire in a lateral valley leading to Hill 942 (3110). But soon several hundred yards of Mount Rosselan's high, vertical rocky walls confronted us and barred further ascent. Encircling the enemy from the south proved to be impossible, leaving a frontal attack against the pass as the alternative.

It took us hours to climb up through the rocks and get at the enemy south of the pass road. The capable riflemen carried the heavy machine guns on their shoulders across places where I had trouble getting along without a pack. Not until just before the fall of darkness did my completely exhausted detachment reach the snow-covered knolls seven hundred yards southeast of the pass and established contact with the units of Gössler's detachment lying at the same height some hundred yards north of the pass road. Dwarf pine bushes concealed my men from the enemy, who occupied a semicircular position on the heights immediately to our front.

I gave the exhausted troops some rest and, with Lieutenant Streicher and some scout squads, examined the possibilities for a surprise night attack against the pass. The night was dark with an overcast sky. It was a good thing that the snow between the low clumps of undergrowth helped somewhat! Crunching it under our feet here and there drew fire from the defenders which made it possible to determine the hostile dispositions.

Sketch 57: The night attack against Klautana Pass.

I managed to locate some suitable machine-gun positions about a hundred yards from the actual gap and somewhat higher than the latter. It took several hours of careful and hard work to prepare our fire support plans for the attack. I used the entire machine-gun company. At the same time the 1st and 3d Companies prepared to attack about three hundred yards from the pass and under the fire support of our machine guns.

All machine guns of the machine-gun company were to open fire at midnight and pin down the enemy in the pass gap for two minutes, and then shift their fire to the enemy on both sides of the gap. The 1st and 3d Companies were to move up to right and left of the gully leading to the pass as soon as the heavy machine guns opened fire and take the pass with hand grenades and the bayonet.

Unfortunately I stayed too long with the fire-support platoons. When their guns opened up I was still on the rocky slope some hundred yards from the two assault companies which, to be sure, should have attacked by themselves, but which I wanted to accompany. I rushed forward and, to my astonishment, found the two companies behind their line of departure. Had the leaders failed, or was it the troops themselves? The two minutes of fire for effect by the machine-gun company had passed. The advance movement of the assault troops was no longer synchronized with the fire of the machine-gun company and the enemy in the pass was no longer pinned down. No wonder that the attack was repulsed with losses after a hard hand-grenade struggle. After the unsuccessful attack I withdrew both companies to the starting position.

I was very angry at this failure of the night attack. It was the first attack since the beginning of the war in which I had failed. Hours of hard work had been in vain. A repetition of the attack during the night seemed hopeless and was beyond my dog-tired troops. Their exertions had been such that they needed rest and food before returning to action and neither of these commodities was available in face of the enemy at an elevation of forty-five hundred feet. I also questioned the advisability of massing large forces near the pass in daylight. For these reasons I broke off the engagement. The 5th Company provided security at the pass, as it had done prior to our arrival, and I took my four companies back into the valley near Pecolat. On the way I reported the failure of the night attack to Major Sprösser, whose command post was in a rock fissure halfway up the slope.

We reached Pecolat before dawn and found the few hovels packed with troops. We camped in the open fields. The pack-animal detachment came up and the cooks soon brewed plenty of hot coffee which certainly hit the right spot. Daylight arrived two hours later and as the sun rose I was called to the telephone to receive the following order:

"The Klautana Pass has been vacated by the enemy. The Rommel detachment marches without delay and joins the Gössler detachment. The Battalion follows through Klaut."

Shortly after daybreak scout squads from the 5th Company had found the pass empty of all hostile elements. The joy of having the enemy yield us such an excellent position without a struggle gave us new strength. The Rommel detachment was soon on the road. After a few hours on the road, we reached the pass and were able to judge the good fire effect of the 1st Machine-Gun Company on the hostile pass position. One of the machine guns had evidently covered the pass road just west of the pass along a stretch of several hundred yards and had inflicted numerous casualties. Bloody bandages on both sides of the road gave eloquent testimony to that effect.

Observations: The night attack of the Rommel detachment on the Klautana Pass failed because the combined fire of the machine-gun company and the advance of the assault companies were not synchronized.

II: Pursuit to Cimolais

The matter-of-fact manner in which the mountain troopers carried their heavy burdens was amazing. Without appreciable rests they had been on the road or in battle for twenty-eight consecutive hours. Twice within this time they had climbed the Klautana Pass, a total difference in elevation of some six thousand feet. We moved downhill with a free gait. The Gössler detachment, as advance guard, had a considerable head start; we caught up with them at noon in Klaut village and moved on. The Gössler detachment ran into the enemy near Il Porto and attacked. A serious battle did not develop for the enemy withdrew to the north. While the Gössler detachment (5th Company, 3d Machine-Gun Company) moved toward Il Porto, the Rommel detachment (1st, 2d, and 3d Companies, 1st Machine-Gun Company) left St. Gottardo as advance guard support for the Württemberg Mountain Battalion which had been reinforced by the 1st Battalion of the 26th Imperial and Royal Rifle Regiment and was moving on Cimolais.

In deployed formation the Rommel detachment followed the retreating enemy on the west edge of the valley toward Cimolais. Initially we moved down the wide valley which narrowed as we approached Cimolais —with rocky cliffs almost sixty-six hundred feet high to right and left, Bushy terrain on both sides of the road concealed our movement from the enemy. Some cyclists under Lieutenant Schöffel, and as much of the detachment staff as could find transportation, acted as a sort of security line ahead of the deployed companies.

It was getting dark when we reached the east bank of the Celina just east of Cimolais. The sandy bed, several hundred yards wide, was almost dry. The enemy seemed to have moved on in the direction of Longarone and the town of Cimolais appeared to be occupied. With the cyclists I crossed the Celina on a broad front. Not a shot was heard. Then Lieutenant Streicher and I rode into Cimolais. The mayor greeted us with extreme politeness. He said everything had been prepared for the German troops and he wanted to give me the key to the town hall. The thought came to my mind that the enemy might have laid a trap.

For security I sent the cyclists some distance farther west down the road leading to Longarone. Then the worn-out Rommel detachment moved in and took up temporary quarters in the southern section of the town. It secured the road toward Longarone and the way toward Fornace Stadion. Our accommodations were good and food was abundant. After the enormous accomplishments of the Rommel detachment, thirty-two hours uninterruptedly in battle or marching without long rests, a few hours of sleep had to suffice to make the riflemen ready for new battles. No one knew what lay ahead of us six miles farther on in the Piave valley.

The staff of the Württemberg Mountain Battalion, the signal company, the Schiellein detachment (4th, 6th Companies, 2d Machine-Gun Company), and the 1st Battalion of the 26th Imperial and Royal Rifle Regiment moved into the northern part of Cimolais. The latter provided security to the north. Night had fallen. The cyclists of the Rommel detachment under Lieutenant Schöffel reported the enemy to be in position digging himself in on the slopes of Mount Lodina (1996) (6600) and Mount Cornetto (1793) (5400). This report was transmitted to the battalion.

The battalion order arriving toward midnight read in part:

"While the 3d Company attacks the enemy west of Cimolais in the morning of November 9, from the west edge of Cimolais, the Rommel detachment (1st, 2d Companies, 1st Machine-Gun Company) encircles the hostile positions west of Cimolais via Mount Lodina (making the ascent before daybreak; similar envelopments by Schiellein's detachment (4th, 6th Companies, 2d Machine-Gun Company) via Mount Cornetto (1792), Mount Certen (1882), Erto, and by Gössler's detachment (5th Company, 3d Machine-Gun Company) via Hill 995, 1483 and Erto."

An ascent by night over rugged, impassable rocky mountains sixty-six hundred feet high (an elevation differential of forty-nine hundred feet) seemed impossible to me in view of the complete exhaustion of my men. Shortly after midnight I went to Major Sprösser and asked him to change the orders. I suggested attacking the enemy west of Cimolais frontally with my entire detachment. Major Sprösser unwillingly changed the order so that only one company of the Rommel detachment had to execute the encirclement via Mount Lodina, while the other companies were at my disposal for the frontal attack.

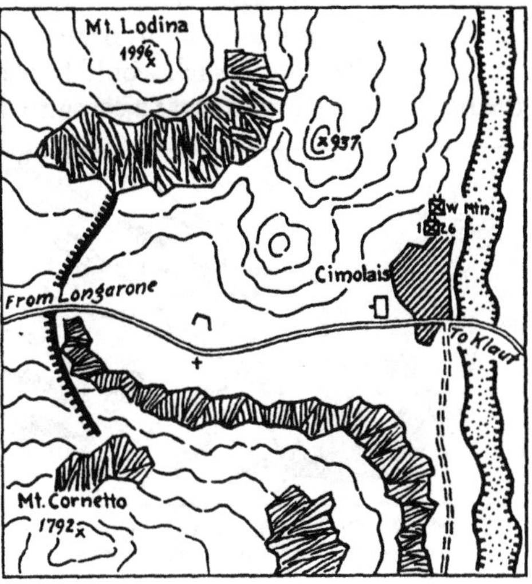

Sketch 58: The situation near Cimolais.

III: Attack Against the Italian Positions West of Cimolais

Three hours before daybreak the 2d Company under Lieutenant Payer, guided by a native, moved off via Mount Lodina to encircle the hostile position to the north. By 0500 Lieutenant Schöffel had determined that the enemy west of Cimolais was completely undisturbed. He assumed that, as on the previous day, the positions had been vacated.

Thereupon I prepared for combat and ordered the mounted company commanders to the southern end of Cimolais. I rode off with my cyclist guards to determine whether the enemy actually had retreated and to reconnoiter the attack terrain in front of the hostile positions on both sides of the pass road. The road rose gently toward the mountain and the cyclists were fifty to a hundred yards ahead of us.

When we reached La Crosett Chapel, 160 yards west of Cimolais, the slopes in front of us began to flash. Machine-gun and rifle fire struck the road and whistled around our ears. In a few seconds the cyclists were off their bicycles and the riders off their horses, which galloped back toward Cimolais. Soon the whole reconnaissance staff gathered in La Crosett Chapel. No one was injured. The walls of the small chapel protected us against the lively fire now concentrated on our shelter. Soon the roof slate began to splinter under Italian machine-gun fire and the fragments cascaded down. The enemy had a better view of his target with each succeeding minute and his nearest position was a bare two hundred yards away. One hostile shell would have been enough to dispatch us into the hereafter. Such a fate was a certainty had we waited.

Sketch 59: The surprise machine-gun attack against the staff on reconnaissance.

When the rifle and machine-gun fire died down a little, I determined the sequence in which we would run back individually from cover to cover toward Cimolais. Sergeant Brückner went first and I followed him. To be sure, the enemy fired heavily on each one of us, but since we ran out in different directions and never left cover at that spot where we had entered, we all succeeded in returning to Cimolais uninjured. Only a few horses were hurt on the reconnaissance trip. All of us would have been killed had the Italians allowed us to continue for an additional hundred yards.

Day had dawned. During the attack the observation squad of the detachment staff under Technical Sergeant Dobelmann had determined the extent of the hostile positions west of Cimolais with the detachment observation telescope (a 40-power glass captured in the Tagliamento incident). The flashes of the discharges in the morning twilight had facilitated their reconnaissance. Dobelmann took me up the Cimolais church tower and showed me the enemy who was in about battalion strength and well entrenched on both sides of the Cimolais-Erto road in fortified and well-developed positions abutting on the vertical rock walls of Mount Lodina about half a mile northwest of Cimolais. The position ran along the steep boulder slope until it crossed the main road six hundred yards west of Cimolais. South of the road it ran along a rock ridge dropping abruptly to the east. The developed and connected position ended 160 yards south of the road. From here the northeast slope of Mount Cornetto was occupied by a hostile skirmish line of about company strength and a few machine guns. The hostile rifleman farthest left was about six hundred yards above the valley floor. The individual riflemen were skillfully dug in with a front toward Cimolais, but the rocky subsoil prevented the preparation of deep positions. Their positions consisted mainly of rocks and stones piled up round about. The hostile positions on the slope of Mount Lodina and on both sides of the road were protected by wire entanglements. The positions on the slope of Mount Cornetto did not need such protection, because vertical rock walls or roof-steep rocky hills made approach almost impossible.

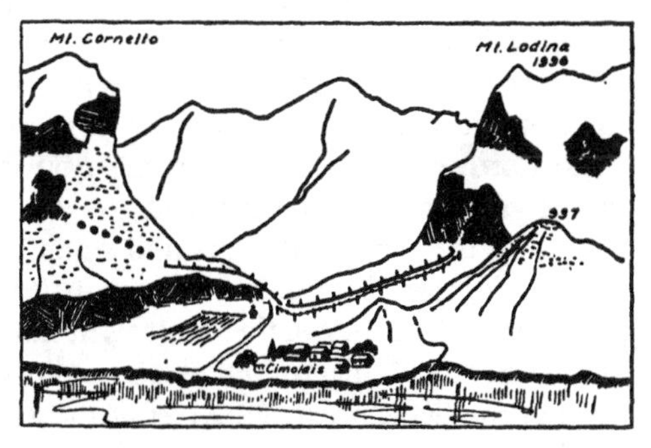

Sketch 60: Positions to the west of Cimolais. View from the east.

During the night I had obligated myself to Major Sprösser to take these positions with a frontal attack. Could I keep my promise? I had imagined the task to be considerably easier. Now it was necessary to make an effort under these difficult conditions. A frontal attack on a broad front and astride the road could be made only against the wire-protected positions on Mount Lodina. It was exposed to flanking fire from Mount Cornetto. Of course it was possible that this effect might be partly counteracted by installing some machine guns on the commanding heights of a foothill of Mount Lodina eight hundred yards north of Cimolais, which the enemy had not included in his position, but the possibilities for adequate fire support for the attack against the wired-in positions were none too bright. An advance against the positions on Mount Cornetto appeared hopeless. A stone avalanche by the defenders would be sufficient to halt an attack without reference to the hostile flanking fire from the Mount Lodina position. Daylight ended the possibility of an envelopment of the hostile positions via Mount Lodina, while a similar undertaking via Mount Cornetto offered no possibility of success. The east slope of the mountain consisted of vertical rock walls which probably no one had ever climbed.

There were no signs of the 2d Company, which had climbed Mount Lodina during the night, and I guessed it had moved to the north and would not be ready to attack until dark. I also figured that Schiellein's and Gössler's enveloping units would not be able to attack prior to that time.

The only location from which suitable supporting fires could be delivered against the enemy positions west of Cimolais was a small hill eight hundred yards north of the town. This was a foothill of Mount Lodina, four thousand feet high, whose summit was covered with low brush. Having thoroughly surveyed the attack terrain with the glass from the church tower of Cimolais, I made the following decision: To attack the garrison on Mount Cornetto with combined fire of several light machine guns from commanding positions on the hill eight hundred yards north of Cimolais, to pin the garrison down, and then to attack up the valley and astride the road.

During the next few hours and unobserved by the enemy, I moved the 1st Company's light machine guns under Lieutenant Triebig into position in the bushes on the knoll eight hundred yards north of Cimolais. I then assembled the gunners and issued my instructions. The remaining units of the detachment (the rest of the 1st Company, 2d Company, 1st Machine-Gun Company) assembled on the covering slopes just northwest of Cimolais and the individual units were given their missions. For the time being no one was committed. The command post was near the 1st Machine-Gun Company; the communications squad established telephone connection with the fire detachment of light machine guns, as well as with the 1st and 3d Companies.

During these preparations, four mountain howitzers and several machine guns of the 1st Battalion of the 26th Imperial and Royal Rifle Regiment opened fire on the Italian pass position from the vicinity of Cimolais church without having previously established contact with the Rommel detachment or made any joint plans. Since this independent firing did not fit in with my plans, I went in person to Major Sprösser's command post in Cimolais and had them cease firing.

At 0900 I ordered the fire detachment of the 1st Company to open fire. According to orders, the fire of four light machine guns covered the hostile riflemen farthest left on the slope of the Cornetto, while two light machine guns pinned down the remainder of the Cornetto garrison. Of course, the distance was excessive for the light machine guns (over fifteen hundred yards), but the effect was excellent. We observed it from various positions with the glass. To be sure, on the southeast flank the Italian riflemen, exposed to the fire from commanding positions, were not hit, but they were so covered with fire that they quickly abandoned their skirmish trenches and sought refuge in the hitherto unendangered zone of their neighbors to the left. The light machine-gun fire of the mountain troops followed them, and it soon became too hot for the Italian soldiers even in their new holes. They moved rapidly toward the prepared position south of the pass road hoping to find shelter there against our fire.

At first only a few Italians moved, but an entire platoon was soon under way. That was what I had been waiting for. The 1st Machine-Gun Company was ordered to take up the fight from the hill just west of Cimolais. Up to that moment we had been unable to occupy that position because of its exposure to fire from the Cornetto. The Cornetto garrison had been driven out. When the first heavy machine guns chimed in, a crowd of Italians (at least a company) on the Cornetto seven hundred yards away rushed panic-stricken toward the southern end of the prepared positions on the cliff 160 yards south of the pass road. The effectiveness of our weapons increased considerably. One heavy machine gun after the other joined the fight. In addition we had the fire of the six light machine guns from very commanding positions. Over yonder men were stampeding toward the narrow trench. It was soon jammed with soldiers and offered little protection against our light machine-gun fire which was exerting most effective plunging fire.

The 3d Company was ordered to attack astride the road. It had nothing to fear from the Cornetto slope, and the machine-gun company was pinning down the remaining Italian positions. The machine guns did their job. While the 3d Company worked its way forward in deep echelon, covered against the fire of the Italian garrison on the Lodina slope, the automatic weapons from the front and above covered the hostile positions south of the road, which were full of men. They pinned down the enemy north of the road and diverted him. The Italian positions south of the road began to empty. Movement was to the rear. The enemy had difficulty in escaping through the close network of German machine-gun fire delivered at 550 yards range. Most of the fleeing enemy were mowed down in a few minutes. I had the conduct of fire under complete control; for I was with the machine-gun company, and had a telephone line to the light machine-gun fire detachment upslope to the left rear.

Sketch 61: The machine-gun attack on the Cornetto garrison.

The 3d Company reached the hostile entanglements and smashed its way into the pass positions, splendidly supported by the heavy and light machine guns. We won!

I ordered the unit of the fire detachment to continue firing. With everything else I followed rapidly into the captured pass position, using the same route as the 3d Company. The hostile garrison on the Lodina slope still held out. A report was made to the battalion regarding the successful attack, and at the same time cyclists, mounted runners, and horses were ordered forward. When I arrived in the captured pass position, the Lodina garrison of two officers and two hundred men also laid down its arms. Particularly pleasing were our few losses; only minor wounds were listed. I had not expected that we could take the hostile position so cheaply.

Units of the hostile garrison fled to the west. My next job was to follow and overtake them and to capture the Piave valley as soon as possible.

Observations: We would have been spared the fire attack of the enemy on the reconnaissance staff had the combat reconnaissance against the enemy west of Cimolais been more thorough during the night of November 8-9.

On the other hand, their fire attack allowed us to determine their precise location. Especially skillful was the utilization of the fire attack by the independent observer of the detachment, Technical Sergeant Dobelmann.

From a technical standpoint the attack at Cimolais was a headache until the precise solution was found. Here the psychological effect of light machine-gun fire even at great distances was taken into account. The first Italian soldiers leaving Mount Cornetto caused panic among their fellows.

The cooperation of the weapons in attack against the enemy west of Cimolais was masterful. Very strong fire was concentrated at the place of breakthrough just before the 3d Company attacked. The well-prepared telephone net made a tight control of the attack possible.

IV: Pursuit Through Erto and Vajont Ravine

We had no time to reorganize, for to leave the fleeing enemy alone, even for a few minutes, might have given his commanders an opportunity to reestablish control. I sent everything I could lay my hands on in pursuit. The rearward units and fire detachments were ordered to move up the road at top speed.

Machine-gun fire from the slopes of Mount Lodina three hundred yards west of the captured positions slowed our pursuit. This fire was coming from elements of our 2d Company who, due to their considerable height, were unable to distinguish friend from foe and thought us to be Italians. We had no protection against their fire and the next few minutes were most disagreeable. Fortunately they realized their error and shifted their fire. During this interlude we lost contact with the enemy and we had to speed up and make up for lost time, for we wanted no more delay short of Longarone. Lieutenant Streicher and I reached St. Martino at 1010 with the leading units of the 3d Company. At the same time the cyclists and mounted runners arrived with the staff's horses from Cimolais.

The road made a very wide curve to the north and debouched half a mile west of St. Martino in the village of Erto-e-Casso. The mountains receded on both sides and small closed columns of Italians were rushing down the road six hundred yards ahead of us. I quickly emplaced a light machine gun for fire support but ordered it to fire only in case we got into a scrap. We chased the enemy down the road. On horse and awheel we soon caught up with the nearest fleeing Italians. No firing ensued. A shout to surrender, a sign to disarm, and an indication of the direction of march which the prisoners were to take was sufficient. We reached and passed through Erto at the gallop. Tethered pack animals stood in the street but there was no shouting. Those we caught up with surrendered without resistance.

Up front with the head of the column, the pursuit looked like a race between horse and bicycle; farther back it looked like the tail end of an army baggage train. Puffing, the soldiers carried their light and heavy machine guns. The Rommel detachment was extended over several miles. Every rifleman realized that it was a question of running down the enemy and that success depended on speed.

The valley narrowed as we approached Erto and the road dropped into the Vajont ravine. Two and a half miles still separated us from our objective—the Piave valley—and the most difficult part of the terrain, namely: the Vajont ravine, still lay ahead. It was two miles long and extraordinarily narrow and deep. Initially the road, blasted into the vertical walls which were four to six hundred feet high, ran along the north wall. A 130 foot bridge crossed the middle of the ravine five hundred feet above the roaring mountain brook. From this bridge the road ran along the south side of the ravine. Various lateral ravines were also bridged and there were several long tunnels along the road proper. A carefully-placed demolition charge would have sufficed to block the road to Longarone for days. As a matter of fact, a machine gun set up at the entrance to a tunnel would have held us up for a long while. All this was to be seen from the map, but I had not had time to study it carefully.

After traversing Erto, the downgrade gave the cyclists a considerable advantage over the horsemen. At a turn of the road they overtook more Italians; then they disappeared from our view. Shortly thereafter shots rang out. Farther forward we saw an Italian automobile driving west. We urged the horses on downhill as fast as they could go, and raced through the first pitch-black tunnel only to be almost thrown from our mounts by a terrific explosion a hundred yards ahead of us. We felt our way through the dark tunnel, which later proved to be full of Italians, toward the outlet. Fifty yards farther we saw the results of the explosion! A deep chasm yawned before us. The enemy had succeeded in blowing up a bridge which spanned a lateral gulch of the Vajont ravine.

Where were my cyclists? Fighting farther west answered my question. We dismounted and I ordered the mounted runner Wörn to bring up all units of the detachment as soon as they came up. Then we climbed to the right in the lateral ravine across the ruins of the blasted bridge and up onto the road again on the other side. We moved forward at a run to the place where the shots were still sounding.

We found the cyclists behind the bridge house at the north end of the bridge which crossed the Vajont ravine in a single span. They were firing on the crew of an Italian truck which had just driven into the tunnel on the other side of the bridge. To all appearances this was a demolition crew left behind to blow all bridges and tunnels which had been previousy prepared for demolition. The cyclists told me they had crossed the other bridge just a few seconds before the explosion, and that Sergeant Fischer had gone up with the bridge in an attempt to pull the smoking fuze from the explosive charge.

Now another bridge lay before us. This one was 130 feet long and five hundred feet above the roaring stream. It was said to be the highest bridge in Italy. Charges were packed into deep square holes in the roadway and we wondered if the fuzes had been lit. The enemy had ceased firing and no longer showed himself at the tunnel mouth. It looked like a withdrawal—if the bridge went up before us, then it would be days before we reached the Piave valley, which was so near. Decisive measures were imperative.

I gave the following order to Sergeant Brückner of the 2d Company, whom I knew to be an especially courageous and reliable soldier: "Take an axe, run across the bridge, and chop through all wires leading to the bridge from that side. As soon as that is done, all of us will follow you in close formation and rip out the fuzes on the way."

A number of low-hanging cables led to the bridge and I was afraid that the Italians might use an electric detonator. The excellent Sergeant Brückner accomplished this mission, and as the last cable broke I rushed forward with the cyclists, tearing out the fuzes on the way. It was in this manner that we gained possession of the undamaged bridge.

We moved on in greatest haste toward the Piave valley. We had to prevent the hostile blasting squad from carrying out its work at defiles along the road. Sergeant Brückner was sent out ahead with a few cyclists. The detachment to the rear received an order to increase its rate of march to the absolute maximum. After passing through several tunnels, the road sloped down to the outlet of the ravine. The vertical wall of rock, into which the road had been blasted, reached a height of fifteen hundred feet. Brückner's squad was not shooting and my guess was that it had reached the mouth of the ravine.

Sketch 62: The Piave River near Longarone.

At 1100 I reached the ravine outlet with some cyclists and riflemen of the 3d Company and of the detachment staff—in all, ten carbines. We were less than a mile from Longarone. It was a beautiful sight. The Piave valley lay before us in the brilliant light of the midday sun. Five hundred feet below us the bright green mountain stream rushed over its broad, multibranched, stony bed. On the far side was Longarone, a long and narrow town; behind it lofty 6,000-foot crags soared up to the heavens. The automobile of the Italian demolition crew was crossing the Piave bridge. An endlessly long hostile column of all arms was marching on the main valley road on the west bank. It was coming from the Dolomites of the north and was heading to the south through Longarone. Longarone and its railway station, as well as Rivalta, were jammed with troops and stalled columns.

V: The Fight at Longarone

The situation in which we found ourselves was one which did not present itself to many soldiers during World War I. Thousands of the enemy, retreating in orderly fashion in a narrow valley hemmed in to the right and left by unscalable mountains sixty-six hundred feet high, were in complete ignorance of the danger which threatened their flank.

Our hearts were young and gay. Those Italian forces must not retreat farther, that was certain. I quickly put my two carbine soldiers in dense clumps of bushes a hundred yards south of the road, and we proceeded to open fire on the columns on the Rivalta-Pirago road at a range of about fourteen hundred yards. We concentrated our fire on a place where escape was impossible for the enemy; on the right, the rocky wall; on the left, the Piave! The leading elements of the 3d Company began to arrive breathless at the mouth of the pass, and they reinforced the firing line.

Sketch 63: The ambush at Longarone.

In a few minutes our rapid fire had split the hostile column in two parts. The northern half marched back toward Longarone while the southern half quickened its pace. Minutes later the enemy turned a large number of machine guns against us. The fire was ineffective for we had occupied good positions among the bushes on the forward slope and had moved away from the debouchment of the road from the Vajont ravine. The Italians only fired at the road and up Vajont ravine, but their actions did slow our advance.

The enemy in Longarone tried to infiltrate to the south. A platoon of the 3d Company, with two light machine guns, was in position south of the Vajont ravine and made life miserable for infiltrators.

Suddenly one of my runners noticed a company of Italian infantry descending the rock walls in our rear (from the direction of Hill 854). I moved a few riflemen and a light machine gun from the firing line to the west to meet this new threat. The enemy continued to climb the steep wall in column of files and came to within three hundred yards of us. Things looked promising, for any man hit when we opened fire would have fallen down the cliff and dragged several of his comrades along. I was sure of success. But I did not fire immediately; rather, I shouted to the enemy to surrender. The enemy saw that the game was up and surrendered. Had we discovered him five minutes later, he would have been behind the steep wall and could have done us much harm.

In the Piave valley the enemy blew up the bridge east of Longarone. An attempt to move off in the direction of Mudu in closed column was thwarted by our fire. Small enemy groups managed to infiltrate south toward Mudu and Belluno. The situation remained unchanged even when several hostile batteries chimed in from the knolls south of Longarone. They did not find our positions south of the Vajont ravine either. Instead, dozens of shells hit on the pass road in front of and in the Vajont ravine, as well as on the cliffs above the road. In spite of the very unpleasant effect of the hostile machine-gun and artillery fire, which was increased considerably by falling rocks and stones, by 1100 the remaining units of the 3d Company, as well as the 1st Company and a platoon of the 1st Machine-gun Company, had reached the heights a hundred yards south of the entrance of the road into the Vajont ravine.

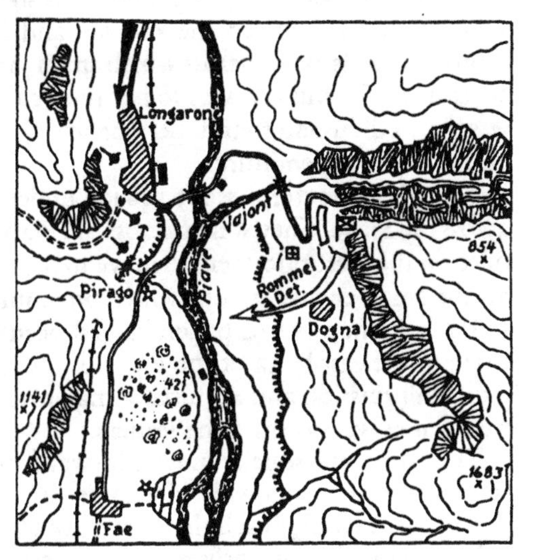

Sketch 64: Crossing the Piave at Pirago and Fae.

In order to block the road and railway toward Belluno on the west bank of the Piave and to capture all hostile units coming from the north, I sent the 1st Company, reinforced by the heavy machine-gun platoon, through Dogna to the west bank of the Piave in the vicinity of Pirago. The entire 3d Company furnished fire support for this movement and prevented the enemy from marching off in closed formation.

In file with very short intervals the 1st Company hurried off in the direction of Dogna. Its way there led over a steep, grassy slope devoid of cover and fully exposed to enemy observation. Italian machine guns and batteries turned their fire on the company, but it managed to reach the protecting houses of Dogna almost without losses. The hostile machine-gun and artillery fire increased perceptibly with its bulk falling in the Vajont ravine.

Then we saw the 1st Company west of Dogna moving up through the bed of the Piave. But the river-bed offered no protection against observation, and still less against fire. Very soon the Italians around Longarone poured such a hail of missles on the 1st Company that only rapid retreat to Dogna prevented heavy losses. While this action was in progress I hurried the detachment staff to Dogna. Telephone wire was laid to the 3d Company, which remained in the former position. Shells and machine-gun fire sped us on our way. The enemy was firing on each individual.

In Dogna I saw the 1st Company, which had just returned from the Piave channel. This failure did not discourage me. That an entire company did not manage to get through the hostile fire zone in the Piave bed did not mean that such a maneuver was impossible for a few men who could take better advantage of the terrain and perhaps veer off somewhat farther to the south.

The heavy machine-gun platoon was so emplaced on the top floor of a house that it covered the railway and highway bridge at Pirago a thousand yards away, across which small groups of Italians were moving southward. Its mission was to bar this road to the larger units. We had less than a thousand rounds per gun and this meant close economy in our ammunition expenditure.

Then I sent several scout squads under particularly capable leaders across the Piave. They were to cross the Piave in very loose formation and, once on the west side, were to move to the vicinity of Pirago and capture all the small enemy groups which were infiltrating south. As soon as a fairly large number of prisoners had been gathered, they were to send them to the east bank of the Piave in the direction of Dogna. This was a difficult mission and required a maximum of skill and dexterity from the men as well as from the leaders.

The five scout squads moved forward under strong fire support but their progress was slow. Under these circumstances, I doubted whether any of them would reach the west bank of the Piave.

Meanwhile Major Sprösser had arrived at the mouth of the pass with the signal company and the attached 1st Battalion of the 26th Imperial and Royal Rifle Regiment. Upon my request the signal company relieved the 3d Company at the mouth of the pass; the 3d Company infiltrated into Dogna.

We saw no signs of the scout squads in the river bed although hostile machine guns were spraying the bare sandbanks of the half mile wide river-bed. Toward 1400 I attacked from Dogna with the 1st and 3d Companies on a broad front in the direction of Pirago. My idea was to get some units across the river and block the valley road on the west side by the fire of the whole detachment. Heavy machine-gun and artillery fire drove us to ground after we had covered a few hundred yards, and we had to dig to get out of the hostile fire. Our net achievement was that we were deployed on a broad front six hundred yards from the enemy's line of retreat and that our attack had drawn the enemy fire from the scout squads farther to the south.

I was skeptical as to whether or not any of the five scout squads had reached the west bank of the Piave, so I sent out the other squads under Lieutenants Streicher and Triebig. The former was soon incapacitated by the blast effect of an Italian shell on the main branch of the Piave, and the latter was wounded by machine-gun fire. It seemed impossible to get a single man across the river. From two sides the Italian artillery cut up the terrain in which we were lying. His guns were in position just south of Longarone and in the vicinity of Mount Degnon (southwest). The enemy did not seem to lack ammunition.

The detachment staff dug in behind a small stone wall in the bed of the Piave. This place was the favorite target of an Italian battery. Various gaps in the stone wall showed that the enemy had made his bracket but we had used the spade to good effect.

Technical Sergeant Dobelmann studied the region south of Longarone with his powerful glass; the adjutant was out on reconnaissance, and I dictated the Cimolais combat report to Sergeant Blattmann who was being trained as detachment clerk. The fire of the enemy continued in undiminished intensity and the 3d Company bore the brunt of it. The enemy continued to infiltrate men and vehicles through the areas covered by our fire.

Toward 1430 the 3d Company and the 1st Machine-Gun Company of the 26th Imperial and Royal Rifle Regiment arrived in Dogna. They were to support us. The leaders reported at my command post. I did not want to expose additional troops to the hostile fire in the river-bed, so I left these new units in reserve in Dogna and committed only one heavy machine-gun platoon, in order to increase the volume of fire being laid by the Württemberg Mountain Battalion on the Longarone-Belluno road and railway. I expected to be across the river before dark.

Seven scout squads had been on their way to the far bank of the Piave for hours. None had reported and I did not know if any had crossed the river. The enemy continued his infiltration to the south and we were powerless to stop him. Ammunition, especially that of the machine guns, was running low and we had to use it sparingly. The minutes dragged and the hostile fire roared on claiming an occasional victim.

Toward 1500 Technical Sergeant Dobelmann reported that he thought he saw mountain troops to the southwest on the opposite slope. He said an Italian coming from the hill west of Fae had been captured by a soldier standing behind a house. I grabbed the glass and convinced myself that everything was in order. No Italians would get past Fae.

But we waited in vain for the agreed return of prisoners to the east bank of the Piave. I had expected to make maximum use of their passage of the river by crossing my people at the same time.

Finally, toward 1530 we saw a dense mass of captured Italians a mile and a half south of us in the broad bed of the Piave. Most of them were already on the east bank heading toward Dogna. I was getting angry because we had lost our chance of shifting to the other bank, when the Italian artillery around Longarone opened on this mass of prisoners. Apparently the artillery thought they were German. The fire forced the prisoners to return to the west bank near Fae. This incident did not change our situation; as before, the enemy kept us pinned down with artillery and machine-gun fire.

Shortly before dark a great number of Italian prisoners appeared near an old levee damming the westernmost arm of the Piave in the vicinity of Hill 431, a mile north of Fae and began to cross the Piave. What I hoped for all day happened. I moved the bulk of my detachment to the weir. We no longer worried about the hostile fire which was still being directed at our old position on the west edge of Dogna.

On the main branch of the Piave hundreds of prisoners protected us from further hostile fire. The shifting of the detachment took little time. The prisoners showed us the best means of crossing the wild river with its many arms, some of which were very swift and chest-deep in places. A single man, even a good swimmer, reached the far shore only with difficulty; the strong current simply carried him away. The Italians grabbed each other's wrists and walked obliquely into the river, facing upstream with the body more or less bent forward, according to the strength of the current. We imitated them and were soon across. However, once there we set out for Fae. The ice-cold bath in the Piave helped maintain a rapid pace.

We were glad to meet the scout squads at Fae. They quickly informed us as to their activities. Deputy Officer Huber and Technical Sergeant Hohnecker with sixteen men of the 1st Company succeeded in fording and swimming the Piave a mile south of Pirago, in spite of the very violent hostile machine-gun fire from Longarone, and took possession of Fae castle. Private Hildebrandt was killed. At Fae the small group blocked the road and railway to Belluno and captured the small squads of Italians coming from Longarone who believed they had reached safety. Lieutenant Schöffel arrived later. In the course of the afternoon the units of the 1st Company at Fae captured 50 Italian officers and 780 men, and an immense quantity of vehicles of all types.

The arrival of reinforcements was most welcome. At times it was rather uncomfortable for so few men to have so many prisoners. Above all the Italian officers needed to be strictly guarded. It had been impossible to move them off, and they were confined in an upper story of the castle and were guarded by two mountain troopers. I had more important things to do than to worry about them.

Our scout squads had cut all telephone lines connecting Longarone and Belluno. I was convinced that help was on the way for the Italians who were trapped in Longarone—at least the hostile battery on Mount Degnon knew exactly what was going on in the vicinity of Longarone. I therefore gave the 3d Company of the 26th Imperial and Royal Rifle Regiment, reinforced by a heavy machine-gun platoon of the Württemberg Mountain Battalion, the mission of providing security and reconnaissance toward the south, the most advanced outpost being about half a mile south of Fae with the reinforced company in the neighborhood of Fae.

I did not count on receiving additional forces. The encircling detachments of the Württemberg Mountain Battalion (the Gössler and Schiellein detachments and the 2d Company), even if they did not encounter enemy, could not have arrived at the mouth of the Vajont ravine, eleven hundred yards east of Longarone, before midnight. There Major Sprösser had the remainder of the 1st Battalion of the 26th Imperial and Royal Rifle Regiment, the signal company of the Württemberg Mountain Battalion and the 377th Mountain Howitzer Detachment, which was out of ammunition.

Should I have been satisfied to block the Piave valley to the north and south on the west bank? Should I have waited until the enemy attacked? No, that was not according to my taste. In order to gain the decision at Longarone quickly, I decided to make a night attack on Longarone with the units of my force still at my disposal (1st, 3d Companies of the Württemberg Mountain Battalion and 1st Machine-Gun Company of the 26th Imperial and Royal Rifle Regiment).

Night fell. The retreat of the enemy from Longarone toward Fae had ceased shortly after our passage of the river. Italian artillery delivered rapid fire in the vicinity of our crossing of the Piave channel. The enemy probably knew that the road to Belluno had been blocked. He had surely seen the eight hundred prisoners and the Rommel detachment crossing from bank to bank in the evening twilight. What did he have up his sleeve? A breakthrough attempt in the night? I had to anticipate him.

By telephone I ordered the heavy machine-gun platoons at Dogna to cease firing on Longarone since we intended to attack that place. Their actions had consisted of delivered harassing fire against all remunerative targets in the vicinity of Longarone and Pirago.

We moved off to the north. I led the point. The column moved in the following order: The light machine gunners marched on the right-hand side of the road with their weapons loaded for steady fire; in the ditch on the left were the riflemen in column of files at ten-yard intervals. The companies followed in columns of files. The detachment staff was at their head. We moved as quietly as possible for hostile sentries had excellent audibility on that clear and quiet night.

In spite of all precautions the point was fired on by an Italian sentry three hundred yards south of Pirago. In the pitch-black night we saw the flashes of a few shots; then my light machine gun on the right hammered away. Its fire struck sparks from the road, a house wall to the right, and the steep rocks on the left of the road. The enemy did not answer for he had been swept away.

We continued the advance, reached Pirago without further encounter with the enemy, and crossed the bridge which we had blocked with fire during the day. Our machine guns at Dogna were silent presumably because of the order sent them by telephone.

We worked our way forward on the road. On the cliff to the left some hundred yards away Italian artillery was firing shell after shell out over us in the direction of the crossing used by us in getting over the Piave. The shell fuzes left a peculiar luminous trail behind them in the dark night. It was a fine gratuitous fireworks display.

Only a hundred yards or so separated us from the first houses of Longarone. We moved forward slowly. There in the light of the fireworks a black wall extended across the bright road. It was about a hundred yards away. We did not know if it was a bend in the road or a roadblock. We moved up to within seventy yards and I made certain that it was a road block. We were expected.

I ordered a halt and brought up the machine-gun company. The company commander (a first lieutenant) received the mission of silently bringing several heavy machine guns into position alongside each other on the road and preparing a fire attack on the barricade. After a short fire for effect I intended to attack with the 1st and 3d Companies and take the south entrance to Longarone.

Preparations for this undertaking were in full swing. The crews of four heavy machine guns were just about to bring their guns into positions eighty yards in front of the barricade, when a sudden burst of machine-gun fire hit us in the flank. Our own machine guns in Dogna were firing. The order to cease fire had not been transmitted to them. Sparks flew everywhere. We tried to take cover and in doing so created quite a rumpus as the machine-gun equipment was banged around. The roadblock opened up, and several machine guns began to sweep the area in which we had taken cover. Machine-gun fire at eighty yards, without a chance of taking cover, is enough to drive one crazy! Death stands very near at such moments. We ourselves did not get a chance to fire. The heavy machine-gun equipment was not assembled. We lay for minutes in the worst kind of crossfire. The attempt to take care of the enemy behind the barricade with hand grenades failed. The distance was too great. To attack on a narrow road against the fire of several machine guns was impossible. We sought shelter under the semicircular recesses in the road wall, and when the fire struck here from the flank, in the ditch to the left. Throwing hand grenades only increased the volume of fire from roadblock. Losses mounted! Among others, the leader of the machine-gun company of the 26th Rifle Regiment lay severely wounded in the ditch to the left. It was a good thing that the night considerably decreased the Italians' accuracy.

The undertaking was a total failure and it remained to get away as quickly as possible without suffering excessive losses. I was pinned down by fire. By word of mouth I transmitted the order to retreat to the bridge near Pirago. The units farther to the rear disengaged themselves with ease but those up front had a more difficult time. Moments when the enemy fire died down were rare. They were taken advantage of for short, quick sprints. We covered a few yards only to be forced to earth again when the machine guns opened up.

A few dashes brought us uninjured to the safety of a bend in the road, safety at least from hostile fire. Unfortunately the machine-gun platoons in Dogna caused a good deal of trouble even here. They blocked the highway bridge at Pirago. I had only a few of my mountain troopers with me. Part were already back in the direction of Pirago, but a considerable number was still up front near the barricade.

Strangely enough, the enemy ceased firing. Shortly after the sound of voices came from that direction and came rapidly nearer. They were not mountain troopers. It was strange that none of the detachment came back. I hurried back to Pirago. On the way I overtook a few mountain soldiers, among them a man with a flare pistol. I found no one at the bridge in Pirago. My order to halt there had not been received.

A group of howling Italians came down the road and I did not know whether they were attackers or prisoners. I had no idea of what had become of my leading elements (3d Company and the machine-gun company of the 26th Rifle Regiment). I decided to use a couple of flares and clear up the situation.

I fired them just to the right of the highway bridge near the low wall leading to the mill and, in their light, I saw a closely packed mass of handkerchief-waving men rushing toward Pirago. The head of the group was a scant hundred yards away and the light of the flares made me an excellent target. The shrieking Italians did not fire a shot as they approached, and I was still undecided regarding their status.

The four or five riflemen with me were insufficient to stem that mob and the rest of the detachment seemed to have gone back in the direction of Fae. I ran down the road intending to catch up with the bulk of my unit, face them about and stop the onrushing horde.

A few minutes later I gathered about fifty men together near a group of houses three to six hundred yards south of Pirago. Lieutenant Streicher took one half of the men and occupied a house on the right of the road, the remainder were used to bar the road. The men lined up with their carbines ready. Lieutenant Schöffel was on the left against the rock wall; Technical Sergeant Dobelmann and I were on the right by the house. The riflemen were instructed to fire only on my command. There were no flare pistols or flare ammunition. The enemy masses were unable to turn off to the left and darkness and lack of time prevented us from determining how things stood on the right where we assumed the Piave to be. We only had a few seconds to complete our preparations. The howling mob drew closer.

Sketch 65: Pirago.

The night limited visibility to a bare fifty yards along the road and the terrain to the right and left was pitch black. When the enemy was within fifty yards I shouted "Halt!" and demanded their surrender. The answering roar was neither affirmative nor negative. No one fired and the yelling mass drew nearer. I repeated my challenge and got the same answer. The Italians opened fire at ten yards. At the same time a salvo rang out on our side, but before we had a chance to reload (light and heavy machine guns were unfortunately missing) we were overwhelmed and trampled down by the powerful mass. Almost all who were on the road fell into the enemy's hands. The bulk of the garrison in the house, whose upper story had only black painted windows, and consequently was poorly adapted for defense, escaped in the dark across the Piave. The Italians raced along the road to the south.

At the last moment I escaped capture by jumping over the road wall and I raced the Italians moving along the road. I tore across country over plowed land, small brooks, over hedges and fences. The 3d Company, 26th Imperial and Royal Rifle Regiment and a heavy machine-gun platoon of the Württemberg Mountain Battalion were still at Fae, a mile away. They faced south and were ignorant of the impending danger. The thought of losing this last remnant of my force gave me superhuman strength. I felt a path under my feet and raced on toward Fae.

I succeeded in arriving before the enemy, and, with everything available, I hastily formed a new front to the north. I was firmly resolved to fight to the last man. Scarcely had the 3d Company of the 26th occupied the north edge of Fae, when we heard the Italians coming down the road. I opened fire when they were still some two to three hundred yards away. The hostile advance slowed down immediately and the Italian machine guns began to rattle, spraying their fire against the walls which sheltered the Styrian troops. The enemy appeared to be attacking to right and left of the road. A thousand men were yelling *"Avanti, avanti!"* ("Forward!")

If I wished to defeat a hostile breakthrough to the south, my reinforced company had to hold a line extending from the sawmill on the Piave four hundred yards east of Fae castle across the north edge of Fae to the cliffs of Mount Degnon three hundred yards west of Fae, or a total front of nearly seven hundred yards. In the middle of this line the reinforced 3d Company of the 26th was already engaged on both sides of the road. Large gaps existed between Fae and the river and Mount Degnon. My last reserves consisted of one or two squads of the 1st and 3d Companies, the remnants of the forces which had advanced against Longarone.

In order to be able to sense hostile attempts at encirclement and in order to have better visibility I ordered a squad of mountain troopers to ignite torches before the entire front from the Piave to Mount Degnon. The riflemen knew that "the chips were down." Soon the sawmill on the Piave was burning and flames began to rise from a large haystack fifty yards to the right of the road and from various houses and barns on the left above the road.

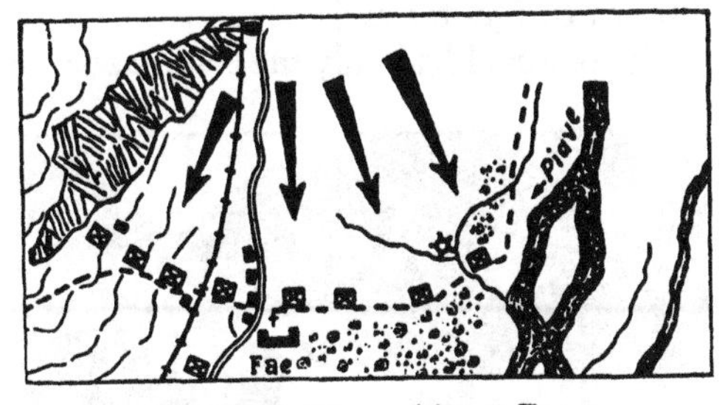

Sketch 66: The position at Fae.

Units of the 3d Company of the 26th were withdrawn from the battle line and used for a continuous, if scant, garrison of the front. In spite of the fierce hostile fire we succeeded in closing all gaps. My valiant orderly Unger offered to get help from the eastern bank of the Piave. He was a good swimmer and thought he had a good chance of getting through. Meanwhile dozens of hostile machine guns hammered against the walls of the castle. The hostile infantry lay densely massed ready for attack about a hundred yards in front of us in ditches and plowed furrows.

Again and again the battle cry: "*Avanti, avanti!*" was heard above the rattle of rifles and machine guns. The rapid fire of the good men of the Styrian and Württemberg Mountain troops prevented the enemy mustering sufficient courage to rise and advance. The enemy's fire front widened.

During this battle, Technical Sergeant Dobelmann, severely wounded, dragged himself across the field in the vicinity of the sawmill and into our lines. The splendid fellow had received a chest wound in the night battle on the road a mile north of Fae, but had been able to escape capture in the darkness and was able to make his way back to us.

I held a few riflemen ready in case the superior enemy succeeded in penetrating our thin line at some place. Two soldiers still held the fifty Italian officers prisoner upstairs in the castle; the latter knowing that their own troops were near at hand became very bellicose, but did not dare attack the two soldiers.

The shots striking the north front of the castle rattled like hell. Most of the Styrians were in position at a wall on the north edge of Fae and fired shot after shot—even if unaimed—over the wall at the enemy. Whenever the Italian fire mounted we increased our fire. This sort of fighting naturally required immense supplies of ammunition. Our supplies would have been soon exhausted had we been unable to fall back on the abundant weapons and stores of ammunition in the castle yard—the booty of the Huber-Hohnecker scouting expedition in the afternoon. In the course of the battle, the rearming of our forward elements with Italian guns and ammunition was accomplished with the help of my few mountain troops. It was none the less unfortunate that the heavy machine-gun platoon in position on both sides of the road had only fifty cartridges for each gun.

For officers I had only the commander of the 3d Company of the 26th and Deputy Officer Huber. All others seemed to have fallen into enemy hands. I missed Lieut. Streicher.

Sketch 67: The Italian night attack.

The battle raged for several hours in undiminished violence. The front between the Piave and Mount Degnon filled and the enemy tried repeatedly to overwhelm us by sheer weight. Our uninterrupted rapid fire prevented a hostile breakthrough to all points. Our southern security element consisted of six men from the 3d Company of the 26th. No others were available. It was already close to midnight. New fires were started on the front, since the old ones threatened to go out. We waited in vain for reinforcements. We believed that units of the 22d Imperial and Royal Infantry Division were arriving on the east bank of the Piave and that the other detachments of the Württemberg Mountain Battalion also were there. We had no telephone communication with Major Sprösser's command post.

The enemy fire slackened shortly after midnight and allowed us to breathe easily. Our losses were moderate, thanks to the skillful use of the small amount of available cover. We worked feverishly to strengthen our positions. Our outposts reported the enemy to be withdrawing and as soon as all firing had ceased we sent out patrols to maintain contact. One of these lost its able leader from fire at close quarters. Another returned at 0100 with 600 prisoners who had surrendered a short distance from our position. The bulk of the enemy withdrew to Longarone.

Reinforcements arrived at 0200 in the form of the entire 2d Company, which under Lieutenant Payer had made the circuit around Mount Lodina, and elements of the 3d and 1st Companies, which, after the night battle south of Pirago, retreated to the east bank of the Piave. We also welcomed the rest of the 1st Machine-Gun Company, which arrived with abundant ammunition, and the 1st and 2d Companies of the 26th Imperial and Royal Rifle Regiment under Captain Kremling.

The whole defense was reorganized and the castle itself converted into a strongpoint. Large quantities of ammunition were available. A company of the 26th Rifle Regiment provided security and reconnaissance to the south. Furthermore, the fifty Italian officers, who had been silent witnesses of the battle at Fae, were sent to the east bank of the Piave. Their crossing of the ice-cold Piave required a great deal of urging on the part of the escort.

The Italians renewed the attack at 0300, but it came as no surprise. They used artillery to support their efforts and we were subjected to a fairly intensive bombardment which did considerable physical damage to the houses at Fae. This preparation was followed by assaults at various points along the front. It turned into a hand to hand fight and we held our strengthened positions and were not even obliged to use any of our reserves. The whole attack lasted a bare twenty minutes and we were left to prepare for the next onslaught.

The Italians had had enough and were satisfied to break off combat and withdraw to Longarone. Their final effort cost them a considerable number of casualties. Unfortunately, the Italian artillery exacted its toll in our ranks.

Shivering, we sat around in our wet clothes and waited for morning. We had located a few bottles of Chianti and we used them to keep warm. Before daybreak the 1st Company reconnoitered the road above the railway as far as the bridge at Pirago. Scout squads of the 2d and 3d Companies reported the territory between the Piave and the Longarone road to be free of the enemy as far north as Pirago. As usual, the scouts returned with prisoners.

By 0630 another battalion of the 26th Imperial and Royal Rifle Regiment had arrived at Fae castle and it was used on security missions to the south. At the same time the Rommel detachment renewed the advance on Longarone. The 2d and 3d Rifle Companies and the 1st Machine-Gun Company marched down the road while the 1st Company moved down the road on the slope above the railway. Our idea was to tighten the noose about the enemy in Longarone.

We met Lieutenant Streicher on the way. He had escaped capture by the Italians in the fight south of Pirago, but in the attempt to cross the Piave was swept several miles down stream and washed ashore unconscious.

The enemy destroyed the Pirago bridges as we approached. On reaching the ruins we found a severely wounded mountain trooper under the debris, but there was no sign of the enemy. We climbed over the wreckage of the iron bridge under the protection of heavy machine guns which had gone into position on the steep slope south of the bridge. As we approached the place on the other side where the roadblock had been the night before, we saw Lieutenant Schöffel on muleback riding towards us from Longarone. Hundreds of handkerchief-waving Italians followed him. Schöffel, who had been captured in the night battle south of Pirago, brought the welcome capitulation of the whole Italian force around Longarone. As drawn up by the Italian commander, it read:

> Longarone
>
> To the commander of the Austrian and German forces:
>
> The troops in Longarone being unable to offer further resistance, this command awaits your decision as to the disposition of our troops.
>
> Major Lay

This happy ending to days of hard fighting made us feel fine, especially since we knew that our comrades, who had been taken prisoner at Pirago, were free again. The Italians lined up on both sides of the road and our march to Longarone was to the accompaniment of their cheer *"Evviva*

Germania!" The commander of the 1st Machine-Gun Company of the 26th Rifle Regiment, who had been captured, severely wounded, by the Italians before Longarone with the greatest part of his company, was driven out toward us in an automobile ambulance. Our progress through the crowded streets was very slow. I went ahead with the ambulance and, in the Longarone marketplace, found the units of my detachment who had been captured. Their arms and equipment had been restored and they held the town pending our arrival. My detachment was the first body of German troops to enter Longarone. We marched in and quartered ourselves in a group of buildings south of the church. It began to rain. There were thousands of Italians and it was slow work moving them from Longarone to the Piave flats to the east. The remaining units of the Württemberg Mountain Battalion, following the 22d Imperial and Royal Infantry Division, marched out of the Vajont ravine.

The other units of the battalion had attempted to come to our assistance during the pursuit and during the fighting on the west bank of the Piave. Immediaely after capture of the Italian positions west of Cimolais, Major Sprösser had initiated pursuit with the signal company of the Württemberg Mountain Battalion and the 1st Battalion of the 26th Imperial and Royal Rifle Regiment. This movement was contrary to the orders of the 43d Infantry Brigade. The nature of the terrain and the type of combat we were engaged in were such that relief by other units was unfeasible. On arrival in St. Martino, Major Sprösser again received orders from the 43d Rifle Brigade:

"The Württemberg Mountain Battalion will halt, camp, and spend the night in the mill at Erto. The 26th Rifle Regiment takes over the advance guard."

Major Sprösser answered:

"The reinforced Württemberg Mountain Battalion is fighting at Longarone and requests infantry support on the pass road and the forwarding of the 377th Imperial and Royal Mountain Howitzer Detachment."

The tenacity with which Major Sprösser stuck to his task, and his refusal to be diverted by orders of the 43d Brigade caused Captain Kremling, the commander of the 1st Battalion of the 26th Imperial and Royal Rifle Regiment, to remark: "I don't know which I should admire more, your courage before the enemy or your courage before your superiors."

Toward noon Major Sprösser reached the exit of the Vajont ravine eleven hundred yards east of Longarone. It took some time for the signal company and units of the 1st Battalion of the 26th to work their way out of the ravine, which was being subjected to heavy enemy fire.

Then the signal company relieved the 3d Company, which was advancing toward Dogna, and fired on the retiring enemy from the heights just south of the debouchment of the Vajont ravine road.

As soon as the leading companies of the 1st Battalion of the 26th Infantry cleared the Vajont ravine at 1400, they were sent to Dogna as reinforcements for the Rommel detachment. No other forces were at Major Sprösser's immediate disposal. The Gössler detachment (5th Company, 3d Machine-Gun Company) had climbed the Forcella Simon (1483) from Il Porto over Cra Ferrona (995). Here its splendid leader, Captain Gössler, an expert mountaineer, had fallen to his death while hurrying ahead of his detachment across an icy slope. The Schiellein detachment (4th and 6th Companies, 2d Machine-Gun Company) had climbed from Fornace Stadion across Mount Gallinut (1303) and had reached the Vajont ravine via Cra Ferrona (995). The 2d Company under Lieutenant Payer was descending Mount Lodina and was headed in the direction of Erto.

After an unsuccessful night attack a series of incredible rumors reached Major Sprösser's command post. One of them was to the effect that the enemy had broken through south of Longarone and had captured me along with the bulk of my detachment. The noise of fighting near Fae soon refuted these rumors.

When our messenger, Private Unger, reached the battalion command post, Major Sprösser sent additional units of the 26th Rifle Regiment through Dogna to Fae, and later the 2d Company, which had arrived from its envelopment of Mount Lodina. The 1st Battalion of the 26th Infantry began to build a footbridge across the Piave west of Dogna.

On November 10, Major Sprösser had his available forces ready for combat on the high ground a thousand yards east of Rivalta. These forces consisted of Schiellein's detachment (4th and 6th Companies and 2d Machine-Gun Company) and the Signal Company of the Württemberg Mountain Battalion, four infantry guns of the 1st Battalion of the 26th Infantry and the 377th Imperial and Royal Mountain Howitzer Detachment. The Grau detachment (5th Company, 3d Machine-Gun Company) was marching up from Erto.

During the night Major Sprösser sent an Italian prisoner of war back to Longarone with the following message written in Italian by Dr. Stemmer:

"Longarone is surrounded by troops of a German-Austrian division. All resistance is useless."

When Major Sprösser found out, at dawn, that the Rommel detachment had renewed its advance against Longarone and that the enemy in Longarone was laying down his arms, he started to march to Longarone with the units of the Württemberg Mountain Battalion located a thousand yards east of Rivalta followed by the 43d Brigade of the 22d Imperial and Royal Rifle Division.

November 10 was a rainy day and it took a long time to clear all Italian soldiers from Longarone's streets. Piles of weapons lay in the public square, and even Italian cannon were delivered there. The lowlands east of Longarone were full of prisoners. In all over 10,000 men—an entire Italian division—had laid down their arms. Our booty amounted to 200 machine guns, 18 mountain cannon, 2 semi-automatic cannon, and more than 600 pack animals, 250 loaded vehicles, 10 trucks, and 2 ambulances.

My detachment's losses in the fighting at Cimolais, in the Vajont ravine, at Dogna, Pirago, and at Fae amounted to 6 dead, 2 severely wounded, 19 slightly wounded, and 1 missing. The losses of the 1st Battalion of the 26th Imperial and Royal Rifle Regiment were unknown.

Lieutenant Schöffel was captured in the attempt to stop the Italians south of Rivalta. At first the Italians slugged him. Upon his complaint, he was brought before a company commander, who did not even apologize for the bad treatment, but wanted to have a personal "souvenir" from the German officer. Then Schöffel had to march along in the front line to Fae. When fighting broke out here, Schöffel lay at the edge of the road close beside an Italian officer who balked all attempts at escape. Schöffel found German fire especially unpleasant. When the Italians disengaged themselves at Fae around midnight, Schöffel was taken back to Longarone, where he encountered the other captured mountain and Styrian troops. Toward morning the captives had to march again toward the south under heavy guard. But they soon halted because, once again, the Italian failed to break through. The prisoners were then led back to Longarone. In the course of the morning, the Italian officers became very friendly toward Schöffel, who made exaggerated statements as to our strength. Finally he was sent toward us with the written message containing the capitulation of the Italian troops in Longarone.

Toward noon on November 10, Longarone was full of German and Austrian troops and sentries with fixed bayonets were required to retain possession of the quarters we had occupied on arrival. Most of my soldiers took off their wet clothes and devoted themselves to well-earned rest in the good and comfortable quarters. In the evening the mountain troops insisted on forming a torchlight procession for their leader.

Observations: After we succeeded in breaking through the hostile positions west of Cimolais, the mobile units (horsemen and cyclists) took over the pursuit of the retreating enemy. They succeeded in catching up with them and, with the exception of one bridge, prevented the Italian demolition squad from doing much damage. This mobile force made it possible to continue the pursuit.

The use of a few riflemen at the ravine exit was sufficient to halt an entire division. The Italians subjected this handful to heavy machine-gun and artillery fire. The riflemen were well dug-in and the fire did little harm. The enemy defensive tactics were incorrect. An attack by part of the enemy forces against the western outlet of the Vajont ravine would have saved the situation.

The attack by the Rommel detachment across the unprotected Piave valley west of Dogna was carried out under heavy fire. The troops made quick use of the spade. Meanwhile weak scout squads on the west bank captured those enemy units escaping to the south after weathering our detachment fire.

Bonfires provided the necessary illumination during the night fighting at Fae, and the ensuing lack of ammunition was made good by rearming with captured Italian guns and ammunition. Both were accomplished under the strongest hostile fire, a remarkable achievement of the mountain troops.

VI: Battles in the Vicinity of Mount Grappa

On orders of the 22d Imperial and Royal Infantry Division, the Württemberg Mountain Battalion moved into the second line and had a day of rest on November 11, 1917, when we buried our dead in the Longarone cemetery.

The momentum of the attack began to slow down. The tempo of pursuit slackened although the enemy was not offering serious resistance in our quarter.

In the course of the following days the mountain troops marched through Belluno to Feltre where they were attached to the German Jäger Division. On November 17 we moved down the Piave from Feltre. Violent fighting was going on in the vicinity of Quero and Mount Tomba and we were soon having difficulty in advancing through the narrow Piave valley, which was jammed with troops. We got into range of the Italian artillery, which was placing heavy interdiction fire on the valley road. Our information was to the effect that the leading Austrian units had encountered strong enemy forces on Mount Tomba.

While in Ciladon we received the mission from division to penetrate the enemy positions as far as Bassano by advancing across Mount Grappa.

In the afternoon the deployed battalion moved into the area just north of Quero, which was under the heaviest Italian artillery fire. The Italian artillery had excellent observation posts on Mount Pallone and Mount

Tomba and it was small wonder that it had registered perfectly on the defile at Quero and on all other important points within range.

Major Sprösser sent the Rommel detachment (2d and 4th Companies, 3d Machine-Gun Company, one-third of the signal company, two mountain batteries, and a radio unit) over Quero—Campo—Uson—Mt. Spinucia—Hills 1268, 1193, to Hill 1306, and the bulk of the Württemberg Mountain Battalion across the Schievenin—Rocca Cisa—Hill 1193 to Hill 1306.

As darkness fell our thin column hastened through Quero at the double. The town had been badly shot up and was still being subjected to Italian artillery fire. Craters of five to ten yards in diameter were no rarity. Large numbers of dead and wounded Jägers lay along our path. Numerous Italian searchlights turned night into day. Simultaneously the heaviest enemy artillery began to strike in the vicinity of Quero, Campo, Uson, and Alano. The searchlights incessantly probed the valley from the direction of Spinucia, Pallone, and Tomba, and the heavy shells roaring up from afar only gave us a few seconds to sprint toward the enemy. During this process contact with both mountain batteries was broken. Sergeant Windbühler was ordered to reestablish it and to bring the batteries up to Uson. The rest of the Rommel detachment succeeded in getting to the village of Uson without losses. The place, like Quero and Campo, was deserted and a ghostly emptiness pervaded all the houses. We were subject to continuous searchlight illumination from the Spinucia and Mount Pallone. The detachment was well dispersed and rested in the shadows of houses and trees. Heavy artillery fire began to come too close for our comfort. Fragments howled through the air, clumps of earth and stones rained down on us, making the bombardment a heavy nervous strain.

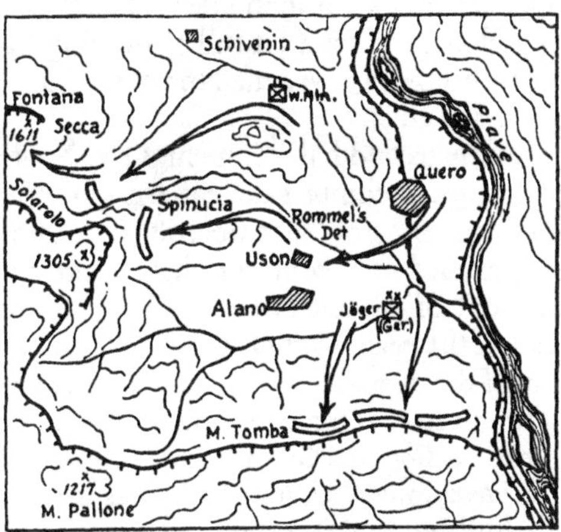

Sketch 68: The advance on Spinucia—Fontana-Secca—Tomba.

Patrols with telephone squads were sent out radially in various directions. Lieutenant Walz took one toward the Spinucia. In my opinion it was no longer a question of a quick penetration across Mount Grappa to Bassano. The enemy front was continuous and strong and we were too late. Six French and five English divisions were said to have come to the Italians' rescue.

The reports began to arrive at midnight. Contact had been established with the neighboring units at Alano. Lieutenant Walz climbed the east spur of Mt. Spinucia without encountering enemy. Sergeant Windbühler brought both mountain batteries to Uson. He had first marched them up the Uson-Ponte-della-Tua valley where they had discovered a well-illuminated barracks. Windbühler halted the batteries, slipped up to the building alone, and found it full of sleeping Italians. Being a fearless man, he drew his pistol, awakened the enemy, and captured 150 soldiers and two machine guns.

In the second half of the night of November 17-18, 1917, the Rommel detachment climbed the eastern spur of Mount Spinucia where, in the early morning of November 18 our leading elements encountered a well-entrenched enemy on the sharp ridge running from the east up to the summit of Mount Spinucia. The positions were less than half a mile east of the peak. Frontal attack without artillery and mortar support was out of the question. The enemy completely covered the sharp, rocky ridge with numerous machine guns deeply echeloned and with mountain batteries on Fontana Secca and Mount Pallone. There was no opportunity for envelopments and it looked as if we were stalled.

Until November 23, 1917, we continued our efforts to advance up the slopes of Mount Spinucia. We were without artillery and mortar support and all our efforts were to no avail. On November 21 Sergeant Paul Martin (6th Company) was killed beside me in an advance observation post by the splinter of an Italian mountain-gun shell. At the same time a Hungarian lieutenant of artillery was severely wounded. On November 23, 1917, the Rommel detachment moved to rejoin the battalion at Rocca Cisa. Here, on November 21, the Füchtner detachment in conjunction with Austrian and Bosnian infantry had attacked and taken the Italian positions on Fontana Secca and on Hill 1222.

At daybreak on November 24, 1917, the entire Württemberg Mountain Battalion was under my command and was disposed in second line on the northeast slope of the Fontana Secca. We constituted the Sprösser Group reserves and had the 1st Imperial Infantry ahead of us. After a successful attack by the Imperials against Mount Solarolo, the Württemberg Mountain Battalion was to smash through in the direction of Mount Grappa. We waited for an Austrian success, standing for hours on the Fontana Secca in snow, ice and bitter cold under the very annoying fire of Italian mountain batteries. The attack against Solarolo made no progress. Our artillery support was too meager and the hostile artillery was too strong. About noon the report came from the Sprösser group that the 25th Imperial and Royal Mountain Brigade had taken Mount Solarolo from the west.

Since the situation on the south slope of the Fontana Secca had not changed at all, the Imperial Rifle Regiment had not moved forward to any extent, and since there was no prospect that things would change during the course of the day, I asked permission to move to the right to the 25th Mountain Brigade in the vicinity of Solarolo and thus to attack in the direction of Mount Grappa. Major Sprösser agreed. Soon the entire Württemberg Mountain Battalion was on the road. It proved possible to take the shortest route, that is, to cross the almost vertical rock walls of the west slope of the Fontana Secca. The alternative was to drop down into the Stizzone valley. We stepped out briskly, but darkness overtook us at Dai Silvestri. I allowed the exhausted Württemberg Mountain Battalion to rest here and sent Lieutenant Ammann (6th Company) to reconnoiter the situation of our troops on Mount Solarolo. My purpose was to march on so early that the rested Württemberg Mountain Battalion would be on the Solarolo and ready to continue the attack at daybreak on November 25. When Lieutenant Ammann returned from his very thorough and successful reconnaissance, the situation had changed. The Württemberg Mountain Battalion was highly censured for moving into the combat territory of the more successful neighboring brigade. Feelings ran so high that Major Sprösser had no other recourse than to ask for immediate detachment from the 22d Imperial and Royal Infantry Division. This was granted. The battalion spent a few days in rest quarters east of Feltre, and then on December 10 moved to the front again down the Piave to the Fontana Secca.

In the night of December 15-16 my detachment bivouacked in snow and ice at forty-three hundred feet elevation. On December 16 the positions on Pyramid Dome, Solarolo (1672) and Star Dome were reconnoitered. The enemy still clung tenaciously to the most important points of these dominating heights. In the night of December 16-17 we were snowed in in our tents. On the next day the Sprösser group attacked. We succeeded in penetrating the positions on the Star Dome, capturing 120 Bersaglieri of the Ravenna Brigade, and repulsing very strong hostile counterattacks. Our own losses, however, were considerable. Quanite, the splendid Sergeant of the 2d Company, did not return from a patrol. Doubtless he was wounded and died of exposure.

In the icy cold we held out on the abrupt slopes of the Star Dome under heavy Italian artillery fire until the evening of December 18, 1917, and then the Württemberg Mountain Battalion began its march to the valley and toward Schievenin. Mail awaited us and there were two small packages among it. They contained the *Pour le Mérite* for Major Sprösser and me. Two awards was a hitherto unheard-of honor for one battalion.

We spent Christmas Eve in small villages north of Feltre. On Christmas Day the mountain soldiers again moved through the narrow Piave valley toward the front under their old Alpino—as the Major was called. My detachment was installed in the Pallone sector with the left flank on Mount Tomba. We had relieved the Prussian infantry. Positions existed in name only. The individual machine-gun and rifle nests were small depressions on the steep, bare slopes and offered little cover. Snow was everywhere! But the cold was still bearable. By day the soldiers had to lie well concealed in their tents, for the enemy had observation over the entire area. Fires could not be built and provisions came up only at night. Tracks in the snow had to be carefully swept out every time. It was too bad when the artillery or mortars aimed at a nest! Some of the companies had dwindled to twenty-five or thirty-five men. In spite of that, they performed their hard and dangerous job with the greatest assurance.

On December 28, 1917, an Italian attack was repulsed on the front of the Württemberg Mountain Battalion. The next day we were subjected to heavy artillery and mortar fire. The Italian heavy mortars were fired at thirty-three hundred yards range and were most unpleasant. On this day the hostile artillery also bombarded the rear area near Alano where the Sprösser staff was located. Gas was used repeatedly.

On December 30, 1917, the enemy increased the violence of his fire on Mount Tomba to the maximum. Enemy planes dove down to within a few feet of our positions and strafed the garrisons with their machine guns. After hours of fighting, the French Alpine troops succeeded in capturing the positions of the 3d Imperial and Royal Mountain Brigade on our left. We held our own, but our left flank was up in the air. A further enemy advance from Tomba in the direction of Alano would have cut us off and obliged us to cut our way out during the night. It was snowing and growing colder!

In the early morning of December 31, reserves moved into the yawning gap on our left. But they suffered heavily from the Italian artillery fire coming from Mount Pallone. The command therefore decided to withdraw the front about a mile and a half to the north. We held our Pallone and Mount Tomba positions until late at night on January 1, 1918. It was bitterly cold. Two of the bravest men fell during the last minutes at the advanced machine-gun post. They were Sergeant Morlok and Private Scheidel. A heavy machine gun jammed while being used against a thirty-man raiding party. Hand-to-hand fighting developed. While part of the garrison held off the numerically superior enemy with pistols and hand grenades, Morlok and Scheidel tried feverishly to get the frozen heavy machine gun back in action. An Italian egg hand grenade fell between them and wounded both mortally. The enemy was beaten off.

Shortly before midnight the Rommel detachment, the rearguard of the Württemberg Mountain Battalion, arrived at Alano with the two victims, and then moved silently up the Piave across fields strewn with dead at Campo and Quero.

A week later I traveled home with Major Sprösser through Trento on leave whence, to my great sorrow, I was not to return to the mountain troops. I was attached to a higher headquarters as assistant staff officer. With heavy heart I followed the course of the Württemberg Mountain Battalion and Regiment during the last year of the war: The great battle in France, the capture of the Chemin des Dames, the attack on Fort Conde, on Chazelle and the Paris position, the battles in the forest of Villers-Cotterets, the crossing of the Marne, the retreat across the Marne, and the battles at Verdun. These battles tore great gaps in the ranks of the victors of Mount Cosna, Kolovrat, Matajur, Cimolais, and Longarone. Only a few of them were destined to see their native land again.

In the east, west, and south are to be found the last resting places of those German soldiers who, for home and country, followed the path of duty to the bitter end. They are a constant reminder to us who remain behind and to our future generations that we must not fail them when it becomes a question of making sacrifices for Germany.

CPSIA information can be obtained
at www.ICGtesting.com
Printed in the USA
LVHW051607281120
672436LV00007B/56

9 781626 54319